brain fuel

brain fuel

199 Mind-Expanding Inquiries *into the*
Science *of* Everyday Life

JOE SCHWARCZ, PhD

Director, McGill University
Office for Science and Society

ANCHOR CANADA

Anchor Canada and colophon are trademarks.

Library and Archives Canada Cataloguing in Publication
has been applied for.

ISBN 978-0-385-66603-9

Cover images: © clipart.com
Book design: Leah Springate
Printed and bound in the USA

Published in Canada by
Anchor Canada, a division of
Random House of Canada Limited

Visit Random House of Canada Limited's website: www.randomhouse.ca

BVG   10   9   8   7   6   5   4   3   2   1

contents

brain fuel

introduction

We humans are a hungry lot. Like all other animals, we of course hunger for food. Unlike our fellow creatures, though, we hunger for something else as well. We hunger for knowledge. Some of this is for practical reasons. We want to know what to eat, what medications to take, what toxic substances to avoid and what to do about climate change. But we also hunger for knowledge just for its own sake. We are innately curious about our history, about the possibility of extraterrestrial life, about why a rose smells like a rose and about why we are curious about so many things. Our brains, like our bodies, constantly need fuel. This book aims to help satisfy that hunger.

This book aims to challenge, too. But, most assuredly, it is not a book of "science trivia." Far from it. Each entry serves a purpose. Some offer serious scientific discussions relevant to daily life;

others are designed to provoke a "Gee, I didn't know that!" reaction. If you are looking for practical consumer information, it's here as well. If you are searching for curious anecdotes to spice up a conversation, you'll find plenty. And if all you want is some personal edification, just keep the book by your bedside, thumb through a few questions every night, and you'll be smarter in the morning! *Brain Fuel* is nutrition for the brain. Digest the whole book and you'll have a pretty good feel for what the pursuit of science is all about.

I also admit to another motive. To me, the pursuit of science is wondrous and satisfying. I of course realize that not everyone shares my passion, and nor does everyone need to, but I do feel that too many are missing out on the benefits that the fulfillment of scientific curiosity can bring to life, and I would like to remedy that. Curiosity, it has been said, is to science what a spark is to a flame. My hope is that at least for some of you, I can kindle that spark into a roaring flame. You will enjoy the internal warmth it provides. I certainly do.

So let's get going. And the best way to get going is to take a look at where we have been. Let's start by going back . . .

potions
from the past

What substance became known as "anaesthesia à la reine" after it was introduced in the nineteenth century?

Chloroform. The "reine" involved was Queen Victoria, the first monarch to give birth to a child under anaesthesia. Prince Leopold, the Queen's eighth child, was born in 1853, after her physician, Dr. John Snow, had administered chloroform by holding a hand-kerchief saturated with the chemical over her majesty's mouth. The results were so satisfactory that the Queen asked for chloroform for her next delivery as well, after which the chemical came to be known in Britain as "anaesthesia à la reine."

Chloroform was first made by the French chemist Jean-Baptiste Dumas, who reacted acetic acid with chlorine, but its use as an anaesthetic was pioneered by James Simpson, a Scottish physician. On the fourth of November, 1847, Simpson and his friends, aware of the euphoria-inducing effects of substances such as laughing gas (nitrous oxide) and ether, sought a little entertainment by inhaling chloroform. After some initial hilarity, they all passed out. Simpson's reaction, on waking, was that "this is far stronger and better than ether." (Ether had been introduced the previous year by William Morton in Boston.) Four days later, Simpson successfully delivered a baby after chloroforming the mother. Within a month he had used chloroform on more than fifty patients, one of whom was so delighted with its effectiveness that she named her newborn daughter Anaesthesia.

The procedure was not without risk, and in 1848 the first death attributed to chloroform was recorded. The death of young Hannah Green was probably caused by improper administration of the anaesthetic. Green's death, along with the Calvinist Church of Scotland's opposition to chloroform, cast a shadow on its use. The Church opposed the use of any anaesthetic in childbirth, reasoning that God had punished all of Eve's descendants by ensuring that women would bring forth children in pain. (It seems that Eve's decision to tempt Adam with that fruit of the tree of knowledge was not a good one.) Opposition to the use of chloroform, however, evaporated when Queen Victoria agreed to be anaesthetized for the birth of Prince Leopold. Approval by the Queen was as close as you could get to approval by God, and the use of chloroform proliferated. Soon it was even incorporated into various patent medicines such as Chlorodyne and Hamlin's Wizard Oil. These "cure-alls" were not only useless but dangerous. Ingestion of significant amounts of chloroform can cause liver damage.

Today, chloroform is no longer used as an anaesthetic, but since it is a by-product of chlorination we are exposed to small doses in our drinking water. Whether or not this presents a lifetime risk is debatable, but chloroform is readily removed by using a home water filter. Bottled waters do not contain any chloroform because they are not treated with chlorine.

The 1905 Nobel Prize in Chemistry was awarded to Adolf von Baeyer for the synthesis of a compound that eventually proved to be important to blue jean manufacturers and also reduced starvation in India. What was this compound?

Indigo—which Adolf von Baeyer synthesized, and determined the exact molecular structure of, in 1880. He was awarded the Nobel Prize by the Royal Swedish Academy of Sciences in 1905 for "his services in the advancement of organic chemistry and the chemical industry, through his work on organic dyes and hydroaromatic compounds." Indigo occurs naturally in the shrubs of the genus *Indigofera* and was well known since antiquity. It was the dye used to colour all sort of fabric blue, including that used to make the uniforms of British sailors. The shrub was cultivated in India on huge farms and exported to Europe. Baeyer's discovery made possible the synthesis of indigo from chemicals readily isolated from coal tar. Levi Strauss used indigo to dye his famous blue jeans. More importantly, the discovery of a process for making synthetic indigo freed up thousands and thousands of acres in India for planting with cereal crops. This fed far more people than the indigo export business ever did.

In 1903 a French chemist dropped a glass flask. It shattered, but the fragments did not fly apart. What had he discovered?

Edouard Benedictus's clumsiness led to the discovery of safety glass. When Benedictus examined the flask he had dropped on the floor, he realized that a film had formed on the inside of the vessel. The flask had contained an alcohol solution of collodion, a plastic made by treating cotton with a mixture of sulphuric and nitric acids. When the solvent evaporated, a film of plastic was left on the inside of the glass. Benedictus thought no more of this until he read a story about a young girl being cut by glass in one of the first automobile accidents. He spent the night trying to make a coating on glass and within a day had produced the first sheet of "safety glass," which he named triplex since it consisted of a sandwich of two sheets of glass with a film of cellulose nitrate between them.

In 1909 Benedictus obtained a patent, and triplex went into production. The first practical use turned out to be in the face shields of gas masks in World War I, but by the 1920s triplex was a standard item in American automobiles. One problem, however, was that the cellulose nitrate yellowed with age. In 1933 triplex was replaced by cellulose acetate, which was not quite as strong but did not yellow. The synthetic resin polyvinyl butyral eventually was found to be superior and has been standard in windshields since 1939. And it all began when a chemist couldn't hold on to a flask.

℘

At the 1937 World's Fair in Paris, the German company Rohm and Haas was awarded a gold medal for a novel material it had produced from acetone. On exhibit was a transparent violin made of the substance. What was it?

Polymethyl methacrylate, better known as Plexiglas. The Rohm and Haas Company of Darmstadt had developed a crystal-clear break-resistant plastic by linking individual methyl methacrylate molecules into long chains of polymethyl methacrylate.

The starting material, methyl methacrylate, was made by reacting acetone with hydrogen cyanide, followed by treatment with sulphuric acid and methanol. The first applications of Plexiglas were for watch glasses and lenses for protective goggles, but soon curved windshields for buses and airplane canopies were being manufactured from the material. Today, Plexiglas has a wide variety of uses, including protecting spectators from flying pucks in hockey arenas.

Why is a 3-foot (1-metre) iron rod on display at the Warren Anatomical Museum in Boston?

Because it passed through the skull of Phineas Gage in 1848 without killing him. It did, however, dramatically alter his personality. The unfortunate event turned out to be a landmark in the history of neurology, demonstrating that different parts of the brain had different functions. Apparently, catastrophic injury to the frontal lobes of the brain could be sustained without causing significant neurological deficits, but not without affecting behaviour.

To this day, a memorial plaque marks the spot where the spectacular accident occurred on September 13, 1848, in Cavendish, Vermont. Gage was employed by a railroad company as a foreman in charge of a crew laying new track. One of his tasks was to blast apart any boulders in the way. This involved boring a hole into the rock and filling it with gunpowder using a long iron tamping rod. On the fateful day, a spark ignited the powder prematurely, propelling the 11-pound (5-kilogram) rod through Gage's left cheek and out the top of his head, landing some distance away. Miraculously he survived, in spite of having lost a significant portion of his brain. Not only did Gage survive but within minutes he was walking and conversing normally. The only immediate consequence was loss of vision in his left eye, which apparently did not prevent him from sitting down and recording the event in his notebook.

Gage's luck, however, did not last long, as he developed an infection that left him comatose for a month. During this time he was looked after by Dr. John Harlow, who skilfully covered the head wound and later recorded the fascinating case in the *Boston Medical Surgery Journal.* In his account Harlow described how the physical injury had altered the victim's personality to the extent that he was "no longer Gage." Although his memory was not affected, the formerly mild-mannered Gage now became capricious and obstinate, often peppering his speech with obscenities. He lost his job and for a while exhibited himself with the famous iron rod at P.T. Barnum's circus.

Gage's most unusual adventure stimulated Scottish neurologist Dr. David Ferrier to investigate the role of the prefrontal lobes in brain function. Ferrier removed the lobes from monkeys and noted that there were no great physiological changes as a result, but the animals' character and behaviour were altered. Today, it is well understood that the prefrontal cortex of the brain controls the organization of behaviour, including emotions and inhibitions.

Phineas Gage died of epilepsy twelve years after the celebrated accident, leaving behind a fascinating legacy and advancing our understanding of the relation between the mind and the brain. Gage's skull has become a relic and is on display along with the famous iron rod at the Warren Anatomical Museum in Boston.

One could say that Gage needed the job with the railway company like he needed a hole in the head.

What substance was responsible for the Chinese ceding Hong Kong to the British in 1842?

Opium. After the discovery of tobacco in the fifteenth century, tobacco smoking became very popular among sailors, who introduced it into China, India, Japan and Siam. In China the practice became so widespread that in 1644 Emperor Tsung Chen prohibited the use of tobacco. The people then turned to opium. By the end of the century one-quarter of the population was using opium. There wasn't enough to go around, though, and the British East India Company began to meet the demand by supplying huge amounts, mostly smuggled into China via Canton by British and American merchants and traded for tea. The Chinese government got fed up and decided to put an end to the illegal trade in 1839, when a thousand tons of opium were seized and destroyed. The port of Canton was closed to the British, who didn't take kindly to this, and the first Opium War was under way. It lasted till 1842, during which time ten thousand British troops captured several ports and the Chinese capitulated. At the peace conference in Nanking, the Chinese ceded Hong Kong to the British and had to concede greater trading rights. Opium flowed into China, and with Chinese immigrants made its way to America and Australia. So it was immoral British behaviour in the early 1800s that eventually resulted in the worldwide opium problem we have today.

Four hundred years ago Belgian physician Johann Baptist Van Helmont was persecuted by the Roman Catholic Church for promoting the use of the powder of sympathy. What was this?

The powder of sympathy was used to treat wounds by applying it to dressings after they had been removed. The exact nature of the substance varied, but iron or copper sulphate seem to have been common ingredients. This silliness was first proposed by physician and scoundrel Sir Kenelm Digby, but Van Helmont bought into the idea. Somehow the effect of the powder on the bloody dressing was to be communicated to the blood still in the body. Why these metal sulphates were supposed to have an effect on the blood at all isn't clear. The Catholic Church held that the powder of sympathy idea promoted superstition and persecuted Van Helmont for his beliefs. Actually, Van Helmont did not believe the practice to be supernatural; he thought it was a perfectly natural phenomenon. Such curious views were not unusual at the time. In fact Paracelsus, who was one of the first physicians to use specific drugs to treat specific diseases, also believed that treating a sword that had caused a wound would help to heal the wound. He described an ointment consisting essentially of the moss from the skull of a man who had died a violent death, combined with boar's and bear's fat, burnt worms, dried boar's brain, red sandalwood and mummy, which was to be applied to the weapon that had inflicted the wound.

The Royal Navy in 1687 tested the notion of sympathetic powder. A dog was wounded and sent off to sea while its bandage remained in London. At a prescribed time the bandage would be treated with the powder and the dog was to feel the effect. Apparently it did not, because the navy did not pursue the practice.

Although his belief in the powder of sympathy tarnishes Van Helmont's scientific reputation, he did make some valuable contributions to science. He was the first to systematically study the production of gases in chemical reactions. He realized that when charcoal burned it released what he called "a wild spirit." This in fact was carbon dioxide. Van Helmont even introduced the word gas into the English language, apparently deriving it from the Greek term for chaos. He studied other gases as well. A red gas, which today we know as nitrogen dioxide, was released when nitric acid, then known as aqua fortis, was poured onto silver. Burning sulphur released sulphur dioxide. Van Helmont even found that intestinal gas was flammable. And he showed that burning gunpowder in a closed vessel caused an explosion because of the production of gases.

In spite of these important findings, we best remember Van Helmont for his classic misinterpretation of his famous tree experiment. Believing that trees were composed of water, he designed an experiment to test the hypothesis. He weighed out exactly 200 pounds (90 kilograms) of earth, moistened it with water and planted a small willow tree weighing 5 pounds (2 kilograms). For five years he diligently watered it and watched the tree grow. Then Helmont weighed the soil, which still weighed the original 200 pounds, and found the tree to be 169 pounds (77 kilograms). He concluded that the extra 164 pounds (75 kilograms) must have come from the added water. The man who had spent so much of his life studying gases never considered the possibility that the tree may have been taking up a gas, such as carbon dioxide, from the air. He had made an interesting observation, but came to the wrong conclusion.

♀

During World War II there was a shortage of wool
because of the large amounts needed to keep
soldiers warm. Researchers at the U.S. Department
of Agriculture developed a fibre called Aralac, which
they hoped could serve as a replacement. What was
it made of?

Milk protein. Natural fibres used to make clothing can be divided
into two categories based on their chemical composition. Cellulosic
fibres derive from plants such as cotton and flax, whereas protein
fibres are made from animal hair such as wool or from insect exudates
like silk. But protein fibres can also be made from virtually any form
of soluble protein.

As early as 1904, Frederich Todtenhaupt, a German chemist,
patented a fibre made from milk, but the company he formed eventually went bankrupt. The process was resurrected by the Italian company SNIA Viscosa, an effort that was supported by Mussolini, who
thought that such a fibre, named Lanital, would help the Italian economy to be self-sufficient. In the meantime, across the ocean, a coop-
erative project between the U.S. National Dairy Corporation and the
Department of Agriculture aimed to find a replacement for wool.
The result was Aralac, made from milk casein. It was widely adver-
tised as a wonderful new fibre derived from milk, the "very fabric of
our lives." Hollywood star Dorothy Lamour was enlisted to model
dresses and hats made from the new material that would take the
sting out of the wool shortage. Not everyone was satisfied with the
garments made from casein, though. Consumers complained that
when the clothes got wet, they smelled like sour milk. And with the
advent of synthetic fibres like nylon and polyester, people went back

to drinking milk instead of wearing it.

But today, fibres made from milk protein are being resuscitated as eco-friendly materials because, unlike the synthetics, they are free of components derived from petroleum. But they are not free of problems. People who have a serious milk allergy may have a reaction to clothing made of "spun cows' milk fibre," especially when they sweat.

In 1775 the French Academy of Sciences offered a prize to anyone who could develop an efficient process for producing a chemical that was much needed for the manufacture of soap and glass. What was that chemical?

Sodium carbonate, also commonly known as soda. To make soap, sodium carbonate is combined with animal or vegetable fat, and to make glass it is heated together with sand. Both these processes have a long history and until the eighteenth century relied on isolating sodium carbonate from ashes left behind when wood or some other plant substance burned. However, as the demand for soap and glass increased, a more abundant source of soda was required, and the French Academy of Sciences offered a prize to anyone who could produce it from salt, which was cheap and readily available.

Nicolas Leblanc, a French physician, took up the challenge, having developed an interest in chemistry while studying medicine. He succeeded in producing sodium carbonate from salt in a two-step process. First, salt was heated with concentrated sulphuric acid, producing sodium sulphate and hydrogen chloride gas. The sodium sulphate was then crushed and heated with charcoal and limestone to yield sodium carbonate. With his patron, the Duke of Orléans, Leblanc established a factory for making sodium carbonate and claimed the prize that had been offered. However, he never did collect. The French Revolution got in the way, and the Duke was accused of "royalism" and was guillotined. Leblanc's factory was taken away and nationalized, and the Committee of Public Safety forced Leblanc to publish details of his process without any compensation. The plant was returned to Leblanc when Napoleon came to power, but no funds were allocated for restoring it to operation. A desolate Nicolas Leblanc committed suicide.

His process, however, was eventually put into production by European chemical plants, making sodium carbonate readily available. The glass and soap industries prospered, as did others like paper manufacturing that had come to rely on soda. In Britain, Leblanc's process provoked one of the first environmental protection acts ever introduced. The hydrogen chloride released in soda manufacture caused problems as it dissolved in rainwater to form hydrochloric acid, resulting in significant acid rain. Britain's Alkali Act required soda manufacturers to pass their effluent gases through acid-absorbing towers. But this did not solve all the environmental problems introduced by the Leblanc process, since calcium sulphide, another by-product, was dumped in fields, where it slowly released toxic and foul-smelling hydrogen sulphide. Today, the Leblanc process has been replaced either by the Solvay process, which is more efficient at making soda from salt, or by isolating soda from mineral deposits known as trona, found abundantly in Wyoming. Still, the Leblanc process retains its historic significance as the first commercial method of making sodium carbonate, an important industrial chemical. Think of it next time you wash yourself with soap, or take a swig out of a glass.

Oats were a common food for the poor in the England of Charles Dickens. The rich, however, never considered eating oats. But they did use the grain for another purpose. What was it?

Pain relief! And it wasn't through eating the oats. They used the grain to make a primitive heating pad. The oats were heated in a frying pan and then placed in a linen bag that was applied to the painful area, usually the stomach. We still make use of this technology. Oats retain heat for a long time and can be effectively warmed up in a microwave oven. That's why oat-filled cloth bags, very much like those used in Dickensian times, are available today to treat various aches and pains.

Of course, eating oats is no longer restricted to the poor. Oliver Twist was fed gruel because it was cheap and easily made. He didn't care about the possible health benefits of eating oats; all he cared about was not being hungry. Oliver even dared to ask for more, and that surely wasn't because he enjoyed the taste. Gruel back then was nothing more than porridge made with crushed oats and water, unlike oatmeal today that's prepared with milk, brown sugar and cinnamon. Oliver Twist may not have had a varied diet, but he undoubtedly had his cholesterol well under control. We now know that beta-glucan, a form of soluble fibre found in oats, lowers blood cholesterol and that a generous daily serving of oatmeal can sometimes be as effective as some of the cholesterol-lowering drugs. You do, however, want to minimize the sugar you dump into it. Try some fruit instead!

"There is a cold draught in this chamber and I request the window be closed." That was the only recorded statement made by one of the world's greatest scientists during the two years he served as a member of the English Parliament at the end of the seventeenth century. Who was he?

Isaac Newton. One of the greatest scientific minds of all time, he was elected twice to the House of Commons, in 1689 and 1701. One would think that the man who formulated physical laws explaining the falling of apples and the motion of the planets would have some clever ideas about legislation as well. Apparently, though, Newton was content with figuring out how tides worked and how a prism separated white light into the colours of the rainbow. He did add his two cents to public service, though, when he served as director of the Royal Mint. Newton's interests apparently did not extend to a social life. There is no record of any romantic involvement during his eighty-four years, and his biographers suggest he died a virgin. But we will never know for sure.

Only once in history has the Nobel Prize in Physics been awarded to a father-and-son team. Who were they, and how did the research for which they received the prize advance the pharmaceutical industry?

William Henry Bragg (1862–1942) and his son, William Lawrence Bragg (1890–1971), who in 1915 were awarded the Nobel Prize in Physics for their seminal roles in X-ray crystallography. This technique involved beaming X-rays onto a crystal of an unknown substance and figuring out the structure of the molecules that make up the crystal by studying the pattern in which the rays are scattered by the atoms in the compound. Chemists were now able to determine the exact molecular structures of compounds. Pharmaceutical chemistry depends on knowing the structure of molecules that make up drugs, and the Braggs' efforts made this determination possible in a relatively easy fashion.

℞

William Harvey's experiments leading to his discovery of the circulation of blood also led to a change in what age-old human practice?

Embalming. Prior to Harvey's demonstration in the seventeenth century of the circulation of the blood, embalming techniques were based upon methodologies developed by the ancient Egyptians. The brain, intestines and vital organs were removed and the body cavities were filled with various aromatic resins and perfumes. Incisions were stitched shut, and the body was immersed in potassium nitrate, usually called saltpetre. Harvey's injection of coloured solutions into the arteries of cadavers suggested the possibility of treating corpses with preservative solutions that would spread throughout the body.

The public first heard about this technique when Martin van Butchell, an eccentric British dentist, had his wife embalmed and displayed in his office to attract potential clients. He had contacted his former teacher, Dr. William Hunter, to do the job. Hunter, together with his younger brother John, managed to preserve Mrs. Butchell, although it is not clear what they injected into her body. Formaldehyde was not yet known, but a solution of potassium nitrate could have been used. They added a dye to the solution to impart a rosy glow to the corpse and replaced her eyes with attractive glass ones before displaying her in a glass coffin. It was rumoured that a clause in their marriage certificate had provided an income for Butchell as long as his wife was "above ground," but there is a good chance that Butchell himself started the rumour to drum up business. The body eventually ended up in John Hunter's medical museum when Butchell remarried and his new wife wanted her predecessor out of the house. The deteriorating body was finally destroyed in a German bombing raid in 1941. We will never know what was in that original embalming fluid.

§

In 1989 the price of palladium temporarily skyrocketed. Why?

Two chemists, Martin Fleischmann and Stanley Pons, working at the University of Utah, stunned the scientific community and the world by announcing that they had discovered the Holy Grail of physics. They claimed to have achieved nuclear fusion in a test tube. Fleischmann and Pons said they had fused deuterium atoms into helium under simple laboratory conditions. The key to the experiment was their use of palladium electrodes. This element was discovered in 1803 by William Hyde Wollaston in a sample of gold ore from Brazil. He first called the shiny metal "new silver" and offered it for sale at six times the price of gold. At the time it was customary to name new elements after heavenly bodies, and palladium derived its name from Pallas, an asteroid discovered in 1802. Palladium could readily be made into jewellery, and unlike silver did not tarnish. Sales, however, were slow. Eventually palladium did find an important use as an excellent catalyst, a substance that can speed up chemical reactions.

Catalytic converters in automobiles make use of palladium's ability to convert carbon monoxide to carbon dioxide and to convert unburned hydrocarbons to carbon dioxide and water. Furthermore, nitric oxide, a contributor to smog, reacts with carbon monoxide in a converter to form carbon dioxide and nitrogen gas.

In 1989, it looked as if palladium was going to reach true glory. The metal has an amazing ability to absorb hydrogen gas and its isotopic relative, deuterium gas. Pons and Fleischmann claimed that when heavy water, or D_2O, was subjected to electrolysis, the deuterium gas produced was absorbed into the platinum electrodes, where it broke down into deuterium atoms, which in turn fused to yield helium with the concomitant production of energy. Critics immediately attacked this proposition and wondered what happened to the neutrons that should have been released by the process. The cold fusion idea has now been pretty well discredited, but there are still proponents toiling away to try to prove that palladium has real magic and will solve our energy problems. Not very likely.

In the 1880s Harold Brown set up a series of "Westinghousing" experiments to which newspaper photographers were invited. What were they supposed to take pictures of?

Stray cats and dogs being electrocuted. In the late 1880s the two giants in the burgeoning field of electricity, namely Thomas Edison and George Westinghouse, squared off in what has been called the Battle of the Currents. Edison was the champion of transmitting electricity by direct current, whereas Westinghouse thought that alternating current was far more efficient since it could transform electricity to higher voltages, which meant a lower loss of transmission over distances. History of course would prove him right, but Edison would have none of it, even though his direct-current method had a problem lighting bulbs as little as half a mile away. Edison was a great inventor, but pity anyone who disagreed with him. The Wizard of Menlo Park could be ruthless, as Westinghouse found out. Edison mounted a campaign to convince people that alternating current was dangerous, and the best way to do this was by publicly demonstrating its lethal effects.

Working behind the scenes, Edison helped his former assistant Harold Brown organize a series of grotesque experiments. Stray dogs and cats were secured onto metal plates that were then connected to a 1,000-volt alternating-current supply. The creatures were then duly "Westinghoused." Edison's real coup, though, came later. To Brown, who had become a member of a New York State commission to investigate more humane alternatives to execution than hanging, he suggested the use of alternating current. He was sure that when the public heard about this, they would associate Westinghouse's current with death, and his direct current would triumph.

Brown executed two calves and a horse by the use of alternating current and convinced the New York Department of Prisons of the effectiveness of the method. It wasn't as bloody as decapitation, as barbaric as hanging or as "pleasant" as injecting a high dose of morphine. The first man sentenced to death by means of the electric chair was William Kemmler, a murderer. The execution on August 6, 1890, did not go smoothly. The victim's death took fifteen horrible minutes. Edison was probably pleased, but he did of course lose the Battle of Currents for the simple reason that he was wrong about direct current being superior for transmitting electricity.

Why did black hair dye for men became a hot seller in the Detroit area in the 1930s?

Henry Ford had decided that younger men were more productive and began to fire older workers from his automobile assembly lines and hire younger ones. This prompted some of the veteran workers to try to disguise their age with black hair dye. While Ford does deserve credit for laying the foundations to the modern automobile industry, particularly by introducing the concept of the assembly line, he also can be roundly criticized for his labour practices and ideological views. He was a blatant anti-Semite and commonly expressed his sentiment that Jews were responsible for many of America's problems. Ford did introduce a five-dollar-a-day plan for his workers, which at the time was more than any other company paid, but his policies constantly harassed workers to be more efficient and to adhere to Ford's lifestyle principles if they wanted to share in the company's profits.

♀

Passover is a celebration of the Jewish exodus from slavery in Egypt. Only after ten plagues had been inflicted on the Egyptians did Pharaoh agree to let the Israelites leave. The first plague was the turning of the waters of the Nile into blood. What scientific explanation has been offered for this?

The most widely quoted theory is that a bloom of toxic algae coloured the waters red and killed the fish in the river. The Old Testament says that Moses struck the water of the Nile, all the water was turned into blood, the fish in the Nile died, and the river smelled so bad the Egyptians could not drink the water (Exodus 7:20–21). The phenomenon of "red tide" is an interesting one. It is caused by the blooming of microscopic single-celled plants called algae and is well known in saltwater seas. Some algae contain a red pigment and can grow so fast that they actually colour the water red. The requirements for such rapid multiplication are long hours of daylight, warm temperatures, and water that is rich in nutrients. These conditions could have been met in the Nile delta. Annual flooding of the Nile was well known before the building of the Aswan Dam, and the flood waters could have delivered nutrients from the soil. Various red tide organisms, such as *Gymnodinium breve*, produce toxins that can kill fish, as the biblical account suggests. Although red tides commonly occur in the ocean, they have been known in river deltas.

Of course, this is just a theory. Maybe there really was divine intervention and the river turned to blood. Or maybe it was divine intervention that caused the red tide. Or maybe the whole thing never happened at all.

§

In 1924 Clarence Birdseye, a former fur trader,
founded the first company to sell frozen foods. Where
did he get the idea for the technology he patented?

From Natives in Labrador who used freezing as a method of
preservation.

Commercial freezing had certainly been tried before, but the
problem always was deterioration of the food's texture. This happened because as water freezes it expands, and the ice crystals ruin
the cellular structure. When Birdseye visited Labrador in 1916 he
ate fish that had been previously frozen but still tasted fine and had
good texture. He realized that the Natives made use of the wind
and cold temperatures to freeze their fish quickly. The freezing was
so quick that the ice crystals that formed were very small and did
not disrupt the texture. By 1924 Birdseye had developed a system
for packing fish, meat or vegetables into waxed cardboard cartons
that were flash-frozen. He founded the General Seafoods Company
and became a wealthy man, changing America's eating habits in the
process. Birdseye went on to introduce refrigerated railway boxcars
and refrigerated glass display cases. Today, frozen foods are widely
consumed and are often more nutritious than fresh. Vegetables, for
example, that are frozen immediately after being picked retain
nutrients that otherwise would be reduced during transport.

The story of frozen foods, though, would not be complete
without a reference to Francis Bacon, the English gentleman scientist who supposedly came up with the original idea. In 1626 Bacon
was travelling through London in his carriage on a rare snowy day
when he began to wonder if cold temperatures might delay the
putrefaction of tissue. He decided to carry out an experiment

immediately and bought a freshly killed hen and stuffed it with snow. Unfortunately he never concluded his experiments because he soon came down with a chill that turned into bronchitis and killed him. This part of the story was always looked on with skepticism because scientific wisdom had it that chills don't cause colds. But recent evidence from the Common Cold Centre in Britain counters this long-held opinion. Volunteers were enlisted to sit with their feet in ice water or in empty buckets. Within five days, 29 percent of those in the icy-water group developed sore throats compared with 10 percent in the control group. It seems people may harbour dormant infections that blossom when immunity is reduced by cold temperatures. So if Bacon had been harbouring a virus before he began his chicken-stuffing experiment, he may really have been done in by the cold temperature.

♀

On his fourth voyage to the New World in 1502, Columbus had to stop in the Caribbean. What prevented him from continuing his exploratory journey?

Shipworms. These creatures, which look like worms but are actually clams, have plagued sailors ever since the first wooden ships were launched into salt water. The ancient Egyptians, Greeks and Romans covered their ships with pitch and tar to protect against the devastation that shipworms can cause. The clams burrow into wood as larvae and then grow to between a fraction of an inch and several feet long. Their nourishment is wood, which they destroy as they bore tunnels through it. Eventually the wood begins to disintegrate, as Columbus learned. The damage to wooden sailing vessels was so extensive that diverse efforts were made to control the nasty clams. Since shipworms can live only in salt water, ships were taken into freshwater rivers. The clams can't stand cold, so sailing into northern waters often helped. But eventually an effective method for covering ships' hulls had to be found, and copper plating fit the bill. Not only did this provide a protective barrier but the copper ions that were released proved to be toxic to the clams.

Today's fibreglass and metal vessels are safe from the ravages of shipworms, but that doesn't mean they are no longer a concern. There are plenty of underwater wooden structures like pier pilings for them to attack. And the wood may look solid on the surface but be like a honeycomb inside; the first sign of shipworm attack is collapse of a pier. Wood used to be protected with creosote, but elimination of this practice because of toxicity meant that many piers were constructed with untreated wood. Chromated copper arsenate was introduced as a replacement preservative but it is also being phased out because it releases traces of arsenic. Copper-based alternatives without chromium or arsenic are being developed.

An interesting question is how this creature manages to survive on a diet of just wood. It does so because of a fascinating symbiotic relationship with a bacterium that lives in the shipworm's gill. *Teredinibacter turnerae* secretes cellulase, which breaks wood down into simple sugars, as well as a nitrogenase enzyme that can convert nitrogen from the air into amino acids. Simple sugars provide a source of energy and amino acids are required for the synthesis of proteins. Scientists are interested in studying the shipworm's enzymes because cellulases have potential applications—for example, in making stonewashed jeans. The cellulase eats away the fabric and releases the dye. So shipworms may have brought the wooden sailing vessel industry to its knees, but they may eventually contribute to the faded knees on expensive jeans.

℘

Spanish researchers discovered that treating the
residues in a pot found in the tomb of King
Tutankhamun yielded syringic acid. What did this
prove about King Tut?

That he liked red wine. The ancient Egyptians were proficient
wine makers. Tombs dating back to 2600 BC are adorned with pictures clearly representing grape cultivation and wine making. It
was also common for wine to be stored in tombs, to supply the
departed in the afterlife. Jars found in tombs have often shown
residues of tartaric acid, a clear indication that a grape product
had been stored in them. But until the recent determination of
syringic acid, nobody knew whether the Egyptians preferred red or
white wine. Now we know that King Tut at least was into red. The
colour of red wine is due mostly to a compound in the anthocyanin
family called malvidin-3-glucoside. Over time, red wine slowly
turns brownish as the anthocyanin molecules react with each other
to produce a giant molecule, or polymer. It was this polymer that
the researchers subjected to analysis. They knew that when it reacts
with potassium hydroxide, the polymer breaks down to yield
syringic acid, which can be identified by using a technique known as
liquid chromatography–mass spectrometry. Indeed, when they subjected the residue to such treatment, they found syringic acid.
That's how they surmised that Tut favoured red wine. The king,
however, was not a poster boy for the supposed health benefits of
red wine. He died at the age of eighteen.

℘

In 1890 ads appeared in British newspapers for "The Carbolic Smoke Ball." What were people supposed to do with it?

Squeeze it and inhale the powder that emerged in order to ward off the flu. The supposed active ingredient was a white crystalline compound called carbolic acid, or phenol, which had been made famous by Dr. Joseph Lister, who had used it on the wounds of his surgical patients to cut down on infections. In 1889 a flu epidemic began to sweep across Europe, decimating the population. No segment was spared; the rich and famous were as vulnerable as the poor. Queen Victoria's grandson succumbed to the terrible disease that doctors were powerless to control. They didn't understand where the ailment came from, commonly ascribing it to "miasmas" or "bad airs." Alcohol, quinine, opium and salicin were all tried as cures but failed to bring relief to the stricken.

The situation was ripe for quackery, and the Carbolic Smoke Ball Company rose to the occasion. It took a kernel of scientific truth, namely that phenol had been useful in preventing gangrene in surgical patients, and blew it up out of context to claim that it could cure the flu. Fearful consumers bought the little rubber ball filled with powder, squeezed it under their nose as described in the instructions and inhaled the cloud that emerged. This would surely keep them from getting sick, they thought. After all, if there were no proof for the effectiveness of the product, the company surely would not have offered a £100 reward—a great deal of money in those days—to "any person who contracts the increasing epidemic influenza, colds or any disease caused by taking cold, after having used the carbolic smoke ball according to the printed instructions."

The ads clearly stated that £1,000 had been deposited in a bank to prove that the company's offer was sincere. Well, they should have been more careful. The company did not think it would ever have to contend with the fury of Mrs. Louisa Carlill and her solicitor husband. Louisa had purchased a ball and had used it as instructed, but still came down with the flu. Now she was angry and wanted her hundred pounds. Mrs. Carlill contacted the company but never got a reply. So she sued. The case went to trial. and the Carbolic Smoke Ball Company offered an interesting defence. The hundred-pound reward was just advertising hype, the company claimed, and "only an idiot would believe such extravagant claims." Judge Hawkins didn't buy this and found for the plaintiff. The company appealed, but the verdict was upheld, establishing the principle of what has since been called the unilateral contract. Through its advertisement, the company had offered consumers a contract, namely protection from the flu or £100. Since the carbolic smoke ball did not protect Mrs. Carlill from the flu, she was owed the money.

Today, the web is filled with companies making claims as exorbitant as those made by the Carbolic Smoke Ball Company and selling products that have about the same chance of curing disease as phenol had of curing the flu. But the companies have learned their lesson and do not offer rewards if their products fail. At most they offer a money-back guarantee, which they seldom have to deliver on because most consumers who have been duped by the latest scam feel too ashamed and just want to forget the whole business.

℘

What momentous chemical event took place in one of the stately homes of England on August 1, 1774?

Joseph Priestley discovered oxygen. Priestley was a minister who had an interest in science but not much money. He needed a patron and found one in the Earl of Shelburne, who had heard about Priestley's success in producing the first soda water and invited him to come to Bowood, his grand country home, to carry out scientific investigations under his patronage. And so it happened that on a sunny day in August 1774, perhaps the most important single experiment in the history of chemistry was performed in a laboratory on the estate of the Earl of Shelburne. Using a magnifying glass, Joseph Priestley focused the sun's rays on a sample of "red calx" (mercuric oxide) and noted that an "air" was given off that was insoluble in water. A candle burned in a spectacular fashion when exposed to the gas, and a mouse became more vigorous when confined to a jar filled with it. Eventually Priestley himself inhaled the gas and remarked: "Who can tell but that, in time, this pure air may become a fashionable article in luxury. Hitherto only two mice and myself have had the privilege of breathing it."

Little did Priestley realize that in fact he, the mice and every other living animal had always been inhaling his newly discovered gas with every breath. Joseph Priestley had isolated pure oxygen. While Priestley was a great experimentalist and observer, he was not an astute interpreter of his experiments. He never recognized oxygen for what it was. Priestley firmly believed that what he had created was "dephlogisticated air." At the time the prevailing opinion was that substances that burned did so because they contained "phlogiston," which during combustion was released into the air. When the air became saturated with phlogiston, it would no longer support combustion. That's why a candle burning inside a closed jar was extinguished. So it made sense to Priestley that his "dephlogisticated air" would be able to take up more phlogiston, and that his candle would burn longer and more brightly.

Soon after his classic experiment, he accompanied the Earl on a trip to the Continent, where Priestley met Antoine Lavoisier, the noted French scientist. Priestley carefully described his mercuric

oxide experiment to Lavoisier, who not only repeated it but inter-
preted it correctly. Lavoisier identified oxygen as an element and
determined that air was composed of two substances, one of which
supported combustion while the other did not. The latter he
named azote, from the Greek for "no life," and it is still the French
term for nitrogen. Interestingly, the Swedish chemist Carl Wilhelm
Scheele had independently isolated oxygen at least a year before
Priestley's discovery but did not publish his work until 1777,
whereas Priestley reported his results immediately. Obviously, it
pays to publish.

A Nobel Prize in Physiology and Medicine was once awarded for which discovery that subsequently turned out to be false?

In 1926 Danish pathologist Johannes Fibiger received the Nobel
Prize in Physiology and Medicine "for his discovery of the *Spiroptera
carcinoma.*" Fibiger had shown for the first time that cancer could be
induced experimentally in laboratory animals and he concluded
that a type of worm ingested by the animals was the specific cause.
As it turned out, the worms were not specifically carcinogenic, but
they did induce irritation in the stomach of Fibiger's animals,
which, coupled with the animals' poor nutritional status, specifically a lack of vitamin A in their diet, did result in cancer.

Fibiger's award has often been ridiculed. Much of the criticism,
however, is unwarranted because, although Fibiger was wrong about
the worms, he wasn't wrong about the possibility of cancer being
induced by an external factor. Epidemiological evidence had previ-

ously indicated that cancers could be induced, but researchers had been unable to demonstrate this in the lab. Percival Pott, a British surgeon, had shown in the eighteenth century that chimney sweeps often succumbed to scrotal cancer because of exposure to coal tar, but the effect could not be reproduced in laboratory animals. Similarly, workers in some chemical industries suffered higher cancer rates, but exposing animals to the same chemicals did not induce cancer. It remained for Fibiger to enter the scene and demonstrate that cancer could indeed be provoked in lab animals.

In 1907, while studying tuberculosis in mice, Fibiger noted the presence of stomach tumours, which upon close examination were found to be infested with worms belonging to the *Spiroptera* family. But when he fed mice the worms or their eggs, he failed to induce cancer. Furthermore, he couldn't find any more mice with such tumours in spite of examining over a thousand animals. But then he discovered a sugar refinery in Copenhagen overrun by mice and found that a number of the animals had tumours with the same worm embedded in them as he had previously noted. The refinery was also infested with cockroaches, which dined on the excreta of the mice. Excreta that contained worm eggs. In the body of the cockroach, the eggs developed into larvae, which infected the mice when they ate the cockroaches. The cockroach served as a vehicle to turn the worm eggs into the disease-causing larvae. Fibiger proved this by feeding infected cockroaches to healthy mice, thereby producing cancerous growths in their stomachs, a result that led him to the erroneous conclusion that the *Spiroptera* worms were carcinogenic. Actually, the animals had been suffering from vitamin A deficiency, which damages cells, and it was the tissue irritation caused by the worms that converted the damaged cells into cancer cells. Any other type of tissue irritation would have had the same result.

In any case, Fibiger's work stimulated research in the area of cancer induction, and before long the Japanese oncologist Katsusaburo

Yamagiwa succeeded where others had failed: he produced skin cancer by rubbing coal tar into rabbits' ears. Basically, then, Fibiger's work stimulated the research that led to the identification of certain chemicals as carcinogens. Today, some five hundred specific chemicals are known to potentially cause cancer. These include formaldehyde, arsenic, PCBs and benzene and can be found in such substances as asbestos, diesel exhaust, soot and certain pesticides. For his role in triggering research into such materials, Fibiger deserves credit, not ridicule.

♀

What substance was supposedly first made by the Phoenicians when they built fires on a beach and rested their pots on blocks of natron?

Glass. Basically, glass is made by heating sand and then allowing it to cool down. Sand, or silicon dioxide, has a highly ordered arrangement of its silicon and oxygen atoms—in other words, it has a well-defined crystal structure. When sand is liquefied by heat and then cooled, this ordered arrangement of atoms is lost, resulting in a more random pattern characteristic of a glass. The sand has been "vitrified." Pure sand has a high melting point and becomes viscous when it melts. The addition of natron, a naturally occurring form of sodium carbonate, has the effect of reducing both the melting point and the viscosity. By 1500 BC the Egyptians were making glass bottles, using natron collected from dry lake beds. Indeed, the word natron, as well as the chemical symbol Na for sodium, derive from the name of Wadi Natrum, the place where the Egyptians sourced the material. The Egyptians had another use for natron, namely in mummification.

Today, glass is essentially made the same way as in antiquity but of course there have been improvements. Until the nineteenth century the only way to make window glass was to blow glass and flatten it. Whirling the disc introduced ripples and thickened the edges. For structural stability it made sense to install the thick portions at the bottom. That's why people used to believe that glass flows, based on the glass in old windows being thicker at the bottom. Studies have since shown that glass does not flow until a temperature of 660°F (350°C) is reached. Today, window glass is made by floating liquid glass on molten tin.

Since glass expands when heated and contracts when cooled, rapid cooling can cause it to crack. The addition of boron oxide to the glass mix greatly reduces expansion and contraction and yields Pyrex, or laboratory glassware.

The way the glass reflects light can be altered by mixing in some lead oxide, resulting in heavy crystal ware. When lead compounds are added to molten quartz, a glass with high density, durability and a special brilliance is formed. Typically, lead crystal contains 24–32 percent lead oxide. If beverages are stored for a long time in such crystals, lead can leach out. The maximum allowable level of lead in drinking water is 50 micrograms per litre, a concentration that can be exceeded in wines kept in crystal decanters. Stored in crystal decanters, port wine can steadily increase its lead concentration from 90 to 4,000 micrograms per litre. Brandy stored for more than five years can have over 20,000 micrograms of lead per litre.

Colours in glass are caused by small amounts of impurities such as cobalt oxide (blue), chromium oxide (green) and iron sulphide (amber).

One final point of interest. Natron is used to make Bavarian pretzels. The dough is dipped into a natron solution to give it its brown colour and distinctive flavour when baked.

♀

What consumer item was advertised in the 1960s on the basis of its "poromeric" properties?

Shoes made of DuPont's novel synthetic leather Corfam. The word *poromeric* was coined by DuPont from *porous* and *polymeric*, implying that the material was composed of giant molecules, or polymers, and that like leather, it was porous, meaning that it allowed air to pass through. By the 1960s the word *polymer* had already entered common lexicon, mostly because of the success of DuPont's flagship synthetic polymeric substance, nylon. The public was amazed by this fascinating material, which had so effectively replaced silk in stockings. DuPont hoped that the success of nylon could be duplicated by finding synthetic substitutes for other natural materials. Leather seemed to be an excellent candidate for replacement because of the concern that supply could not meet demand. Furthermore, leather production is quite complicated, with all the chemicals needed for tanning and dyeing, and leather of course also wears out. On the other hand, leather is supple and "breathes" well, making it an excellent material for shoes.

DuPont researchers focused on developing a material that would be as porous as leather but more water resistant, longer lasting and easier to maintain. They came up with Corfam, which incorporated layers of vinyl and polyurethane and was even more porous than leather. On top of that, it could readily be cleaned with a moist cloth and its shine easily restored. Policemen were enlisted to try the newfangled shoes made of Corfam and were delighted with its properties. Their dirty walking shoes could be easily rinsed under running water. Based on such responses, as well as on laboratory tests showing impressive air porosity and water impermeability, DuPont confidently introduced

Corfam to the world in 1963 at the Chicago Shoe Show and featured the new material prominently at the 1964 New York World's Fair. Shoe manufacturers lined up to purchase Corfam, and some seventy-five million pairs of shoes were quickly sold.

But the initial success of Corfam withered. People complained of sweaty feet and lack of comfort. Perhaps the biggest issue, though, was that consumers did not want shoes that lasted for ever. They wanted to keep in step with changing styles. By 1970 people were calling Corfam "DuPont's Edsel," after the notorious failed car brand. With profits waning, DuPont gave up on Corfam and sold its production facilities. That did not end the production of "poromeric" shoes, which are still available from the companies that purchased the DuPont plants and are still favoured by the military, where they make spit shines easier.

☙

In 1982 Germany issued a stamp to commemorate the 100th anniversary of the death of famed chemist Friedrich Wöhler. What molecule is depicted on the stamp?

Urea. Why would a compound found in urine merit such an exalted status? Because in 1828 Friedrich Wöhler made an accidental but momentous discovery that involved urea and changed the face of chemistry.

At the time most scientists believed that substances produced by living species could not be duplicated in the laboratory. Such natural substances, often referred to as organic, were thought to possess a "vital force" that was outside the scope of human intervention.

Urea was certainly an organic substance, having been isolated from urine by the French chemist Hilaire-Marin Rouelle in 1773, and therefore, according to current thought, it could not be made in the laboratory. But in 1828 Wöhler slew this theory. While heating a sample of a clearly inorganic substance, ammonium cyanate, he made a critical observation. The crystals of the cyanate transformed into a novel substance that on analysis proved to be urea. Wöhler excitedly wrote to his former mentor, the Swedish chemist Berzelius: "I must tell you that I can make urea without the use of kidneys, either man or dog. Ammonium cyanate is urea." Well, ammonium cyanate wasn't urea, but its atoms had rearranged to form urea.

Ammonium cyanate and urea were actually isomers, meaning they were composed of the same atoms but joined together in a different fashion. What Wöhler had shown was that the properties of a substance were just a reflection of their chemical composition and no "vital force" was involved. The only requirement for the synthesis of organic compounds in the laboratory was chemical ingenuity. And chemists rose to the challenge. It wasn't long before Emil Fischer was synthesizing proteins and Paul Ehrlich was making antibacterial compounds. Today, chemists can even synthesize pieces of DNA, a molecule that of course has also appeared on many stamps.

♀

On October 4, 1957, the Soviet Union launched *Sputnik 1,* the first man-made satellite to orbit the earth. How did the Soviets prove that they had actually carried out this mission?

Sputnik carried a radio transmitter that generated a constant beeping sound that could be heard by radio astronomers as the satellite passed overhead. Thousands of amateur radio operators as well as, of course, military personnel heard the telltale beep as the satellite orbited the planet every ninety-six minutes. The Space Age had begun!

The stated goal of launching a satellite was to further scientific knowledge and celebrate the International Geophysical Year. But there was more to *Sputnik.* It was launched using a booster rocket that had originally been designed to deliver a nuclear bomb, and now the Soviets had clearly demonstrated their ability to deliver such a weapon across the ocean. Putting a satellite into orbit required a rocket that could achieve an altitude of at least a hundred miles (160 kilometres) and reach a final velocity of at least 17,500 miles per hour (28,000 kilometres per hour). Such a rocket could easily target the United States, which is why there was great concern in American military circles. The situation was deemed to be very serious, especially given that the first two attempts by the U.S. to launch a satellite turned out to be disastrous. Finally, about four months later, the U.S. managed to launch *Explorer 1* successfully, which actually did result in a scientific accomplishment, the discovery of the Van Allen radiation belts. By this time, though, the Soviets had launched *Sputnik 2*, which carried the famous space dog, Laika. The space race was now on in full, eventually culminating in Neil Armstrong's landing on the moon. Indeed, the small step that had been *Sputnik* now, with the Apollo landing, had become a giant leap for mankind.

In ancient China, what were officers of the court required to carry in their mouth when addressing the Emperor?

Cloves. The idea was to protect the exalted ruler from the perturbing effects of bad breath. Cloves, which are the dried buds of a tropical tree, have a pleasant, characteristic fragrance that can mask unpleasant odours. And not only the ones we describe as halitosis. Spoiled food, which of course was common before refrigeration, often has a telltale aroma that can be countered by the addition of cloves. It isn't surprising, then, that one of the reasons explorers like Columbus and Vasco da Gama set sail was a quest for spices. The name *clove* derives from the French *clou* for nails, as the shape of the bud does vaguely resemble a small nail. Chinese courtesans who were unfortunate enough to have a toothache when they sought an audience with the Emperor apparently discovered that cloves had a medicinal effect. The spice has anaesthetic properties. Indeed, its active ingredient, eugenol, is still used today in ointments designed to afford temporary relief from toothache, and dentists commonly swab gums with eugenol before stabbing them with a needle.

Cloves' potent smell made the spice a much-desired commodity when the plague devastated Europe. The disease was believed to be spread by breathing "miasmas," or bad air, which cloves were supposedly able to purify. Aristocrats and doctors armed themselves with pomanders made by poking cloves into dried oranges in an attempt to ward off the plague, a practice that gave rise to the tradition of using potpourri to scent houses. But cloves developed a reputation that extended beyond plague prevention. Many people believed in the doctrine of signatures, which maintained that God had given humans clues about treating diseases with natural substances. Jaundice, for example, was to be treated with yellow turmeric. And since cloves are the only spice shaped like the most characteristic feature of the male anatomy, they were believed to have aphrodisiac properties. Perhaps this explains why at one time cloves were literally worth their weight in gold. Men were looking to spice up more than their food.

food stuff

What common food frightened Alfred Hitchcock so much that he claimed to have never even tasted it?

Egg! You might think that the portly director feared heart disease and was frightened by the cholesterol in the yolk of egg. But that was not the case. He just didn't like the appearance of the yolk. Here is what he told interviewer Oriana Fallaci in 1963: "I'm frightened of eggs, worse than frightened; they revolt me. That white round thing without any holes, and when you break it, inside there's that yellow thing, round, without any holes . . . Brrr! Have you ever seen anything more revolting than an egg yolk breaking and spilling its yellow liquid? Blood is jolly, red. But egg yolk is yellow, revolting. I've never tasted it."

Hitchcock was not a scientist. Otherwise he would have known that eggs do have holes, lots of them. The shell of an egg is made mostly of calcium carbonate, with small amounts of calcium phosphate, magnesium carbonate and proteins blended in. And it is perforated by some nine thousand holes through which moisture and gases can readily pass. Proof of this is that as an egg ages it loses moisture and the contents shrink. As that happens, an air pocket develops, changing the density of the egg. A fresh egg will sink when placed in water, but one that is rather older will float.

The porosity of the shell also allows carbon dioxide to pass out of the egg, which affects both the appearance of the white of the egg and the ease of peeling a boiled egg. The acidity of an egg is controlled mostly by the amount of carbonic acid it contains. Carbonic acid forms when carbon dioxide dissolves in water. As carbon dioxide evaporates through the egg shell, the contents become less acidic, and the proteins that tend to cluster together to make the egg white cloudy now separate from each other, giving a clearer appearance.

The yellow colour of the yolk that revolted Alfred Hitchcock so much comes from pigments called xanthophylls, which are found in the leaves of most plants. The xanthophylls that make their way into the bodies of animals (in the case of humans, in the eye) come in the most part from their food intake. The xanthophyslls in eggs derive from the alfalfa and corn the hens are fed. Sometimes marigold petals are added to the hens' diet to make the yolks even more frighteningly yellow.

Unlike Alfred Hitchcock, most people are more worried about a yolk's cholesterol content than its colour.

An average yolk contains about 250 milligrams of cholesterol, which is a large chunk of the daily 300 milligrams that health experts suggest we consume at most. But the fact is that dietary cholesterol makes a very small contribution to our blood cholesterol levels. Every 100 milligrams of cholesterol intake increases blood cholesterol by about 0.06 millimoles per litre, which is practically insignificant. Indeed, no studies have reported a significant relation between either egg consumption or dietary cholesterol intake and heart disease.

✺

Why does a teaspoon of kosher salt taste saltier on the tongue than a teaspoon of table salt?

It has to do with surface area. Crystals of kosher salt are larger than those of regular table salt, allowing for greater contact with the taste buds. A rabbi does not "bless" the salt to make it kosher (although Morton's Coarse Kosher Salt in the past has claimed to be packaged under rabbinical supervision). As with any other salt, some commercial kosher salt uses anti-caking additives to make it free flowing, but it usually does not contain iodide, which is added to table salt to protect people from thyroid problems. The size and shape of the crystals of kosher salt allow it to absorb more moisture than other forms of salt, and this makes it excellent for curing meats. This is where the name comes from. The salt itself is not kosher, meaning it doesn't conform to Jewish food laws, but the salt is used to make meat kosher. The Jewish holy book, the Torah, prohibits consumption of any blood, which is why kosher meat must come from animals slaughtered in a specific fashion and the meat must be treated to rid it of blood. A common way of removing the final traces of blood from meat is to rub it with salt.

✺

White chocolate has a shelf life of only a few weeks at room temperature, while regular chocolate lasts many months. Why?

Regular chocolate contains plenty of naturally occurring antioxidants known as polyphenols, which are absent from white chocolate. Actually, "white chocolate" isn't really chocolate, because it does not contain the cacao bean particles that are responsible for the characteristic taste of chocolate. All chocolate originates from cacao beans, which are found inside the fruit of the cacao tree. The beans are removed together with the pulp that surrounds them and are piled for a few days to allow fermentation to occur. During this process, yeasts and bacteria that are naturally present in the beans convert sugars to alcohol and acetic acid, which in turn initiate reactions that produce flavourful compounds. After drying, the beans are roasted, and the heat triggers yet more reactions that produce the complex molecules that together constitute the taste of chocolate. The roasted beans are then ground to produce a thick slurry known as chocolate liquor, which when pressed is separated into cocoa cake and cocoa butter. Pulverized cocoa cake gives us cocoa powder, which is used to make the cocoa we drink.

Eating chocolate is made by combining the chocolate liquor with various amounts of cocoa butter, sugar, milk solids and often vanilla for flavouring. Lecithin is commonly added as an emulsifier to prevent the separation of the cocoa fat. White "chocolate" is a concoction of cocoa butter, sugar and vanilla. Since it contains no chocolate liquor, it has no antioxidants that can prevent the fat from going rancid. Rancidity occurs when fats react with oxygen and break down to produce various nasty-smelling compounds like butyric acid. Fats that contain double bonds, in other words unsaturated fats, are more prone to this process. Most of the fat in chocolate is of the saturated variety, with stearic acid being dominant. That's why even white chocolate without its protective polyphenols can last for weeks.

Saturated fats have been implicated in heart disease, but stearic acid converts to oleic acid rapidly in the body. This is the same fatty acid found in olives, and is not linked to heart disease. Furthermore, the antioxidant phenolics that account for as much as 8 percent of cocoa powder also offer health benefits. Of course the real reason to eat chocolate is because it tastes so good. And contrary to what some may think, it is not addictive. Research has shown that "cravings" can be satisfied by imitation chocolate with no cocoa in it, while swallowing tasteless capsules that do contain chocolate leaves chocoholics in a state of frustration.

It supposedly lowers blood pressure, reduces cholesterol and keeps vampires away. What is it?

Garlic. They sure say a lot of things about garlic. It kills bacteria. It energizes. It prevents cancer. It lowers blood cholesterol. Of course there is a big difference between what "they" say and what science says.

What science now says is that garlic *doesn't* lower cholesterol. Cholesterol has become a fearsome word. In people's minds it is often linked with heart disease. Garlic has developed a reputation for lowering cholesterol and therefore the risk of heart disease—but where did this reputation come from? Animal studies, by and large. Over 110 studies in animals have investigated the effect of garlic on blood cholesterol, and about 85 percent of these showed a lowering of cholesterol. That was impressive enough to stimulate human studies, and by 1995 a number of these also indicated that garlic could lower cholesterol. But as bigger and better studies were mounted, the cholesterol-lowering effect seemed to evaporate. The problem was inconsistency in these studies. Some used fresh garlic, some used aged garlic, while others used a plethora of supplements. It isn't surprising that the results were variable because garlic is chemically very complex. Crushing or processing the cloves unleashes a cascade of chemical reactions so that different supplements can have very different chemical profiles. Dr. Christopher Gardner and his group at the Stanford Prevention Research Center in California decided to try to settle the garlic-cholesterol question by designing a bigger and better study than anyone had previously carried out.

One hundred and ninety-two men and women aged between thirty and sixty-five with elevated LDL, commonly called "bad" cholesterol, were divided into three groups. One group would ingest one large clove of fresh garlic in a sandwich daily while the other two groups would take one of two garlic supplements. The supposed active ingredient in garlic is allicin, the compound that forms when the clove is disturbed or processed in some way. The supplements were chosen because they contained allicin to the same extent as the dose of fresh garlic that was eaten. And two supplements were used in the study because the allicin in each case had been processed in a different way. Participants consumed the garlic products for six days a week over a period of six months and had their blood drawn periodically. No differences were seen among the groups; none had any significant cholesterol lowering. This was a large sample, and a well-done study. If garlic had a cholesterol-lowering effect, it would have been seen here.

Of course, this doesn't mean that garlic can't prevent cancer or keep vampires away. Some other well-controlled studies are needed . . .

What fruit added to a spinach salad can significantly increase the absorption of carotenoids like beta-carotene and lutein that are present in the spinach?

Avocado. Carotenoids are effective antioxidants, but they are far more soluble in fat than in water. Unless they are eaten along with some fat, the body absorbs very little of them. The oil in salad dressing helps, of course, but the avocado is a great choice because it is rich in fats but also contributes antioxidants of its own. In a study of eleven men and women, adding half an avocado to an undressed salad dramatically increased carotenoid absorption. Lutein absorption went up fourfold, whereas beta-carotene absorption increased twelve times. Furthermore, the fat in avocado is of the heart-healthy monounsaturated variety.

❦

Today, food additives are carefully regulated, but this was not always the case. Before laws were introduced, chemicals such as copper sulphate, aluminum sulphate and indigo were commonly added to foods. What was their purpose?

To give food a more appealing colour. Copper sulphate was added to pickles to make them more green, aluminum sulphate was used to whiten bread, and indigo was added to tea to darken it. That's the same indigo that was used to colour fabrics, the same one that became famous as the colour of blue jeans.

Just because indigo is used to colour fabrics does not mean it is dangerous in foods, but when used in high doses it can be toxic.

The real problem was that these colourants were used to make poor-quality products appear better. Adding indigo to tea meant that less tea could be used to achieve the same colour. Today, all sorts of food dyes are used, often to make junk food look more appealing. The difference is that dyes today have to pass toxicity tests. Still, as a general rule, foods that contain dyes as additives tend to be of poorer nutritional quality.

♀

What berry, named after a furry bird, contains more vitamin C per pound than oranges and more potassium than bananas?

A bit of a trick question. It's the kiwifruit.

Yes, according to the botanical definition, it is a berry, a berry being a many-seeded pulp fruit, the seeds of which when mature are scattered through the pulp. So by this definition, the strawberry is not a berry. Kiwi is actually the most nutrient-rich of the top twenty-six fruits consumed in the world. In addition to vitamin C and potassium, it is also a good source of folic acid, vitamin E, lutein and dietary fibre. It is also rich in antioxidants that can prevent damage to our DNA. Research attests to the health benefits of kiwis. The berry's protective properties have been demonstrated in a study with six- and seven-year-old children in northern and central Italy. The more kiwi or citrus fruit these children consumed, the less likely they were to have respiratory-related health problems, including wheezing, shortness of breath or night coughing.

✿

What flavour of ice cream is linked to paper-industry by-products?

Vanilla. Natural vanilla flavouring is derived by extracting the vanilla bean with alcohol. Although the extract contains dozens of different compounds, there is one, called vanillin, that is responsible for most of the flavour. This compound can also be readily synthesized. The starting material for the synthesis used to be lignin, the material that is removed from pulp when it is converted into paper. A plant at Thorold, Ontario, used to produce 60 percent of the world supply of synthetic vanillin. Today, most synthetic vanillin is produced from raw materials found in petroleum. For reasons of cost, far more synthetic vanilla than natural is used in ice cream production.

✿

Why are researchers looking into harlequin bugs as a possible food for humans?

Harlequin bugs feed on cruciferous vegetables such as broccoli, cabbage and horseradish. These vegetables have aroused a lot of interest because they contain glucosinolates, which have anti-cancer properties. The bugs concentrate the glucosinolates in their bodies; they can have twenty to thirty times more in their flesh than in the sap in their intestine. We can think of them as little anti-cancer pills.

Why are they called harlequin bugs? Because their black and red colouring makes them look like the medieval pranksters who were known as harlequins. Unfortunately, chances are that harlequin bugs would not taste very good. The glucosinolates are quite bitter. That's why birds who try to feast on harlequins quickly spit them out. But then again, birds don't know about the potential benefits.

☙

In Homer's *Odyssey*, what is characterized as "liquid gold"?

Olive oil. In antiquity it was used to flavour food. It was also the oil Greek athletes used to coat their bodies before they competed naked in the original Olympic Games. Today, people tend to coat the insides of their bodies with it, hoping to avail themselves of its health benefits. In the 1980s studies began to show that LDL, or the "bad" cholesterol, could be lowered if olive oil replaced other fats in the diet. Bertolli, capitalizing on this notion, came up with the slogan "Eat well, live long, be happy." When the Mediterranean diet began to be linked with good health, olive oil again rose in status.

While it is true that monounsaturated fats such as olive oil do not raise LDL cholesterol, they are not unique in this regard. Diets that incorporate canola oil or sunflower oil also fall into this category. But "extra virgin" olive oil may have an added benefit. This is the oil that is obtained by cold-pressing high-quality olives and it contains a variety of antioxidants. These, specifically the polyphenols, may protect against heart disease and cancer. Some epidemiological evidence has indeed shown that people who consume a lot of olive oil may have a lower risk of breast and colon cancer and a lower risk of heart attack. Lower-grade olive oils contain fewer polyphenols, since these tend to be destroyed during processing. Such processing involves taking the mash from pressed olives and extracting the oil with a solvent to remove off-flavours. The solvent—hexane is most commonly used—is then evaporated by heating the oil. Unfortunately the heat also destroys some of the flavour compounds and the polyphenols.

You actually cannot tell the quality of an oil by its colour or price. Taste studies have shown that cheap extra virgin oils are as good as expensive ones, and surveys have demonstrated that light-coloured oils can be of very high quality.

Oenologists hate 2,4,6-trichloroanisole. Why?

Because oenologists are wine makers, and they fear the off-flavours that can be produced by "cork taint." Cork taint is the prime reason people send back bottles of wine. The compound responsible for it is 2,4,6-trichloroanisole, which is produced by fungi that grow on cork trees. Cork is produced from the bark of the tree, and the musty-smelling anisole can form at any stage of cork production by the action of the fungus on chlorophenols, which occur naturally in wood. Humans have an incredible ability to detect "corked" flavour in wine. In fact, we can detect trichloroanisole at a concentration of five parts per trillion. That's like being able to pinpoint one specific second in about a million years. Roughly 15 percent of wine bottles develop this taint, so it is a big expense for wine producers. The answer? Plastic corks or screw caps. That's one reason we're seeing an increase in such closures.

♀

What crop is capable of producing the greatest amount of protein per unit of land?

The soybean. Not only is the soybean high in protein but the type of protein it contains has an excellent amino acid profile, meaning that it contains good doses of all the essential amino acids animals and humans need to survive. Soybean protein can even reduce cholesterol levels in the blood. In Italy, 25 grams of soy protein a day are commonly prescribed to people who have high cholesterol. Soybeans also contain isoflavones, compounds that chemically resemble estrogen. Some menopausal women find that their symptoms are reduced when they consume soy products. There is also intriguing evidence that consuming soy products around the age of puberty reduces the risk of breast cancer in women, although the jury is out on the effect of soy on existing breast cancer. In men, soy intake has been linked with a reduced risk of prostate cancer. Soy oil is low in saturated fats, which makes it beneficial to health, and it can also be converted into various industrial oils and even into biodiesel.

For all these reasons, demand for soy is increasing, and farmers have to meet the challenge. They have certainly seen increased yields by planting what have come to be known as Roundup Ready soybeans. Weeds of course are a great enemy of crops because they suck nutrients from the soil. Roundup, or chemically speaking, glyphosate, is a very effective brand of herbicide. Not only is it effective but it has an extremely low toxicity in mammals, birds and fish. But if it is sprayed on a field of soybeans, it kills the crop as well as the weeds. Enter biotechnology. Scientists found a way to introduce a gene into the soybean that makes it resistant to glyphosate. A field can be sprayed and the weeds destroyed without affecting the crop. This means using more glyphosate, but far less of more toxic herbicides. As well, less tillage is required, meaning less soil erosion and less fuel used for tractors. Is there a "but"? Of course. There always is. Pollen can drift from the glyphosate-resistant soybeans to some weeds, making them resistant. There have been some episodes of this, but so far it has not developed into a problem. The benefits outweigh the risks. So next time you eat that tofu burger, think of all the interesting science behind it.

♀

What are "love apples"?

There are two answers to this one. Firstly: tomatoes. The fruit originated in South America and was introduced to Europeans by the conquistadores in the seventeenth century, but it failed to enthuse European palates. At least until some clever marketer labelled the tomato as an aphrodisiac and started referring to it by the Latin expression *poma amoris.* But it was when the French got into the game and christened the tomato *pomme d'amour* that tomato sales in Europe really took off. Maybe that's the origin of our expression "she's a hot tomato." The French were perceived to be great lovers, and if they thought the tomato had special properties, well then, it surely must be the case. Americans apparently either did not hear of this connection or were just too worried about the safety of tomatoes to be charmed by its aphrodisiac potential. They were convinced that tomatoes, belonging to the same plant family as belladonna, were poisonous. So Americans did not start eating tomatoes until around 1820, when, according to legend, Robert

Gibbon Johnson, in what appeared to many to be a public display
of bravery, ate a tomato in front of a crowd in Salem, New Jersey.

Tomatoes of course do not have aphrodisiac properties, but
they still do deserve our love. They have some pretty interesting
health properties. The fruit is rich in lycopene and vitamin C;
lycopene has been strongly linked with a reduced risk of prostate
and cervical cancer.

Answer number two is that in the Elizabethan Age lovers
exchanged "love apples," which really were apples. But specially
treated apples. A woman would keep a peeled apple in her armpit
until it was saturated with her sweat, and then give it to her sweet-
heart. And guess what? Modern research has shown that there may
well be compounds in underarm secretions that have some attractive properties for members of the opposite sex.

♀

What do people do who have a condition that derives its name from the Latin word for "magpie"?

Eat things that are generally not considered to be food. The condition is known as pica, which is the Latin word for "magpie," a
bird that is thought not to be very discriminating in its food
selection. The most common type of pica, known as geophagia, is
the consumption of earth, especially clay. Other types include
eating laundry starch (amylophagia), hair (trichophagia), raw
potatoes (geomelophagia) or unusual amounts of lettuce (lec-
tophagia). People have also been known to eat paint, plaster,
chalk and paper. Perhaps the most dangerous form of pica, especially for children, is plumbophagia, the ingestion of lead-based

paint. Pregnant women are especially prone to pica and often chew on ice (pagophagia).

Nobody really knows what causes pica. Clay eating may have a historical connection because of its absorptive properties. Ancient physicians used a form of clay know as terra sigillata to treat suspected poisonings because it has the ability to absorb toxins. Indeed, in developing countries where hookworm is a common ailment, clay is used to bind this intestinal parasite. Diarrhea and gas are also sometimes treated with clay, as evidenced by the use of Kaopectate, which has clay as an ingredient. There is also a theory that pica is linked to some sort of nutritional deficiency, particularly that of iron. In some cases iron supplements alleviate the condition, but the studies are not compelling.

Basically, pica remains a mystery. Of course, why people eat Twinkies or poutine is also a mystery.

φ

What edible oil has recently been found to have anti-inflammatory effects similar to ibuprofen, a common anti-inflammatory medication?

Extra virgin olive oil. A compound called oleocanthal has been isolated from extra virgin olive oil and in laboratory studies has been shown to have the same effects as ibuprofen. This discovery came about in a fascinating way. Gary Beauchamp of the Monell Chemical Senses Center in Philadelphia was attending a scientific meeting in Italy where he tried some freshly pressed extra virgin olive oil. The experience wasn't altogether a pleasurable one, as almost immediately he began to feel a stinging sensation in his throat. As luck would have it, Beauchamp had previously worked on testing the sensory properties of ibuprofen medications and had experienced exactly the same effect. Could there be some connection between olive oil and ibuprofen? the Monell scientist wondered. Oleocanthal was eventually isolated from olive oil and was determined to be responsible for the stinging effect. Curiously, though, this compound had no chemical resemblance to ibuprofen. Yet there seemed no doubt that oleocanthal was the "stinger," since a synthetic version added to corn oil resulted in throat irritation. Further experiments revealed that oleocanthal blocked the action of the COX-I and COX-2 enzymes, just like ibuprofen. These enzymes are known to produce inflammation.

The next question to answer was just how much oleocanthal there is in olive oil and how much of the oil would have to be consumed to get an effective dose. The answer is, a great deal! A whole glass of olive oil would be needed to treat a headache. This of course is not recommended, but incorporating some extra virgin olive oil into the diet is a good idea. Many diseases, ranging from dementia and heart disease to some cancers, have been linked to low-grade inflammation, and it just may be that the reason these conditions are less prevalent in Mediterranean countries has something to do with the large amount of olive oil consumed. It may also have a lot to do with the fact that they eat less meat and eat more fruits, vegetables and nuts.

Fill in the blank in the following sentence written over 220 years ago by Thomas Jefferson to John Adams. "The superiority of _____ both for health and nourishment, will soon give it the same preference over tea and coffee in America which it has in Spain."

The missing word is *chocolate*. Jefferson turned out to be wrong: hot chocolate doesn't rival coffee or tea in terms of consumption, but maybe it should! Modern research shows that the warm liquid extract of cocoa beans, introduced to Cortez by Montezuma in 1519, has some interesting health properties. Would you believe a blood-pressure-lowering effect attributable to consuming chocolate? That is just what Jeffrey Blumberg of Tufts University and his colleagues in Italy discovered. Dark chocolate lowered blood pressure by a clinically significant amount Blumberg and his colleagues at the University of L'Aquila, including senior author Dr. Claudio Ferri, studied ten men and ten women, all of whom had hypertension and a systolic blood pressure between 140 and 159 millimetres of mercury (mm Hg) and a diastolic blood pressure (bottom number) between 90 and 99. None of the participants was taking antihypertensive medicines, none had diabetes or other disease, and nor did they smoke.

For one week before starting the study, participants avoided all chocolate and other flavonoid-rich foods. Flavonoids are compounds in many plant products that are believed to have healthful properties. During the next fifteen days, half the group ate a daily 3.5-ounce (100-gram) bar of flavonoid-rich dark chocolate, while the other half ate the same amount of white chocolate. After another week of avoiding flavonoid-rich foods, each subject "crossed over" and ate the other chocolate. White chocolate, which contains no flavonoids, is the perfect control food because it contains all the other ingredients and calories found in dark chocolate.

The researchers found a 12 mm Hg decrease in systolic blood pressure and a 9 mm Hg decrease in diastolic blood pressure in the dark chocolate group after fifteen days. Blood pressure did not decrease in the white chocolate group. This is the kind of reduction in blood pressure often found with other healthful dietary interventions.

The researchers report that the dark chocolate group also experienced a significant reduction in several measures of insulin resistance compared with the white chocolate group. Levels of LDL ("bad") cholesterol dropped by about 10 percent in the dark chocolate group, but stayed the same in the white chocolate group. Surveys show that 52 percent of North American adults say that chocolate is their favourite flavour, so it certainly is easier to convince them to eat more chocolate than, let's say, broccoli. But remember that the key is *dark* chocolate. . The bottom line is that in the context of a balanced diet, 3.5 ounces (100 grams) of dark chocolate a day is not a bad idea.

$$\mathcal{G}$$

The "noble rot" was first recognized in Hungary in the seventeenth century. It has been used ever since. For what purpose?

To make sweet or dessert wines. Grapes can be attacked by various moulds, which usually causes them to rot. That's why vintners use a variety of fungicides to protect the fruit. But infection with one particular mould, *Botrytis cinerea*, can be used to advantage. This was discovered around 1650 in the Tokaj region of Hungary and gave birth to the famous wines named after that region. "Noble rot" perforates the skin of the grape, which causes it to lose moisture and thereby concentrates the sugars. The mould also produces a variety of compounds that add to the taste and alter the body of the wine. Glycerol helps create a full body, while compounds like sotolon and octenol help produce a distinctive flavour. The French recognized a good thing when they saw it and adopted the methods used to produce wine by means of the noble rot, which they termed *la pourriture noble*. If you sample a wine from the Sauternes region of Bordeaux, you'll discover that some rotten things can be pretty good.

A study published in 2003 with the title "Phenolic Compound Contents in Edible Parts of Broccoli Inflorescences after Domestic Cooking" was widely reported in the lay press. What was the surprising finding in this study?

That microwave cooking of broccoli resulted in a 97 percent loss of flavonoids, compounds with antioxidant activity. The paper, published in the *Journal of the Science of Food and Agriculture*, described a study in which researchers cooked broccoli by boiling, steaming or microwaving and then examined nutritional losses. Broccoli was chosen because of its reputation as a "healthy" vegetable, a reputation based on its content of two compounds, namely sulphoraphane and indole-3-carbinol, which have anti-cancer properties. Curiously, these were not the compounds the researchers monitored in the study. Instead they looked at various flavonoids, which are also supposedly beneficial because of their antioxidant properties. Microwave cooking resulted in a 97 percent loss of flavonoids as well as significant losses in other antioxidants. Steaming resulted in minimal losses.

One problem with the study is that too much water was used during microwaving. The researchers used two-thirds of a cup of water for one and a half stalks of broccoli, whereas the usual amount is just one or two tablespoons. This could have resulted in excessive leaching of nutrients. The cooking time was also longer than it needs to be; one to two minutes is sufficient. For some reason nobody has attempted to repeat this work using more realistic conditions.

Microwaves work by heating up water, and since water is distributed throughout broccoli, it is theoretically possible that nutrients are exposed to more heat than during steaming, when heat has to travel from the surface of the florets to the interior. Based upon this one study—and it is always dangerous to conclude too much from one study—steaming is the best way to cook broccoli. Of course, what is really important is to make broccoli—raw, steamed or microwaved—a regular part of the diet.

❦

Why is meat from the breast of a chicken white and from the legs dark?

Dark meat has more myoglobin, an oxygen-storage compound. Contrary to what people may think, the difference is not due to the tissues containing more or less blood. After slaughter, blood is drained from a chicken's carcass, so whatever colour differences appear in the flesh have nothing to do with blood. Rather, the key is the presence of myoglobin. Since chickens don't fly, they don't use their breast muscles much. But when they run around they certainly exercise their leg muscles. Any time a muscle does work, it requires a supply of oxygen, which is provided by hemoglobin in the circulating blood. However, muscles that are used a lot can't get enough oxygen from hemoglobin and rely on another oxygen-storage molecule in muscle called myoglobin. Like hemoglobin, this is red when it carries oxygen and becomes purple after it has given up oxygen. Basically, then, leg muscles exercise more, need more myoglobin and therefore are darker in colour. Game birds have dark breast meat because flying requires more extensive use of breast muscles than clucking about a chicken coop.

☙

What major change occurred in the production of M&M candies in 1976?

The red ones were eliminated. Much to the public's dismay, the Mars and Murray Company stopped production of red M&M's because of a health scare concerning Red Dye Number 2, which at the time was the most commonly used red food dye. It was never used in M&M's, but the company decided to withdraw the red candies "to avoid consumer confusion and concern." It isn't clear exactly what confusion Mars and Murray was worried about, since the U.S. Food and Drug Administration banned Red Dye Number 2 in January of 1976. So if red M&M's had stayed on the market, it would have clearly meant that they didn't contain the suspect dye. Perhaps the company was concerned that people might think it was using an illegal dye.

The story becomes even more bizarre when you examine the evidence upon which the ban was made. In the early 1970s a couple of questionable Soviet studies suggested Red Dye Number 2 caused thyroid tumours in male rats and stillbirths and deformities in female ones. These were followed by some flawed American studies, which even if correct would have implied that a human would have to drink 7,500 cans of coloured soda a day to reach the levels of dye that had been given the rats. Canada was unconvinced by the American studies and never banned Red Dye Number 2. Various rumours began to float about why the red dye was really banned, with the most popular one suggesting that it was an unapproved aphrodisiac. Mars and Murray never addressed this issue, anticipating the eventual return of the red candies. This happened in 1988, after the furor about the toxicity of Red Dye Number 2 had died away. With great fanfare the red M&M's were reintroduced, with some ingenious advertising hinting at their supposed aphrodisiac properties.

$$☿$$

In 1965 a chemist working on an anti-ulcer drug licked his finger when picking up a piece of paper. He was astounded by the taste. What had he discovered?

The artificial sweetener aspartame. It was December 1965, and Jim Schlatter, a chemist at G. D. Searle, was working on a project to discover treatments for gastric ulcers. To test new anti-ulcer drugs, he synthesized some peptides (combinations of amino acids) normally found in the stomach. In the course of his work, Schlatter accidentally got a small amount of one of these compounds on his hands without noticing it. Later that morning, he licked his finger as he reached for a piece of paper, and noticed a sweet taste. His curiosity drove him to ask, "Where did that sweet taste come from?" His first thought was of the doughnut he had eaten during his coffee break, but then he remembered he had washed his hands since then. The sweetness could only have come from the aspartylphenylalanine methyl ester he had worked with. He knew that aspartic acid and phenylalanine, which make up this product, are natural amino acids present in all proteins, so he felt it would be safe to taste the material. It was sweet. He and his lab partner, Harman Lowrie, both tasted the material in 2 teaspoons (10 millilitres) of black coffee, noting the sweet taste as well as the absence of any bitter aftertaste, and recorded the results in Schlatter's laboratory notebook. His boss, Dr. Bob Mazur, convinced the company of the potential value of this discovery. Within twenty years Schlatter's curiosity had turned into a billion-dollar-a-year business

🔋

What is "liberty cabbage"?

Sauerkraut. During World War I, President Woodrow Wilson whipped up a great deal of anti-German sentiment in the United States. Because sauerkraut was a German word, it was renamed liberty cabbage, the same way that after 9/11 french fries were renamed freedom fries. Sauerkraut is fermented cabbage, meaning that bacteria have converted some of the sugars in the cabbage to lactic acid. This acid is a good preservative and also gives the cabbage a characteristic taste. Sauerkraut is an excellent source of vitamin C, as well as of lactobacilli, the bacteria that are found in yogurt. Capt. James Cook always took a store of sauerkraut on his sea voyages, since experience had taught him that it was an effective remedy against scurvy.

The downside of sauerkraut is its high salt content. The salting of the cabbage serves two major purposes. First, through osmosis, it causes the release of water and nutrients from the cabbage leaves. The fluid expelled is an excellent growth medium for the micro-organisms involved in the fermentation. Second, the salt concentration inhibits the growth of many spoilage organisms and pathogens but does not affect lactic acid bacteria. As cabbage is approximately 90 percent water and the salt is dissolved entirely in the water, the actual salt concentration experienced by the micro-organisms in their aquatic milieu is around 2.8 percent. Thorough and even distribution of the salt is critical. Pockets of low or high salt concentration would result in spoilage, lack of the desired fermentation or both. Oxygen must be excluded during the fermentation because its presence would permit the growth of some spoilage organisms, particularly the acid-loving moulds and yeasts.

♀

What health claim has the U.S. Food and Drug Administration allowed barley producers to make?

Manufacturers can claim that whole-grain barley and barley-containing foods that provide at least three-fourths of a gram of soluble fibre per serving can lower the risk of heart disease. The claim is based on the established fact that soluble fibre can reduce blood cholesterol. Fibre is the indigestible component of grains, fruits and vegetables and passes through the digestive tract mostly unchanged. That, though, does not mean that it has no physiological effects. Soluble fibre, so called because it dissolves in water, binds to bile acids in the digestive tract and removes them from the body. Bile acids, which are needed to digest food, are made in the liver and are reabsorbed into the bloodstream after they have done their job and are recycled. But soluble fibre throws a wrench into the works and prevents the bile acids from being reabsorbed. The liver is forced to make more, and guess what the raw material is that the liver requires for this process? Cholesterol! So as cholesterol is converted into bile acids, blood levels of the substance decrease.

The particular soluble fibre that is found in barley is beta-glucan. Studies have shown that 3 grams of this a day can have a significant effect on blood cholesterol levels. The FDA examined five clinical trials that had investigated the result of consuming whole-grain barley and dry-milled barley products and concluded that there was a consistent lowering of blood cholesterol levels. Other grains, such as oats, also contain beta-glucan, but barley has an advantage because its beta-glucan soluble fibre is found through-out the entire barley kernel, rather than only in the outer bran layer. Processing therefore does not remove the beta-glucan, which means that even refined products such as barley flour, barley flakes or bar-ley meal contain beta-glucan.

Of course, the benefits of soluble fibre become meaningful only in the context of a healthy diet. That's why the health claim has to include the statement that soluble fibre in barley can reduce the risk of heart disease *in conjunction with* a diet low in saturated fat and cho-lesterol. So don't get the idea that having bean and barley soup before eating steak and french fries will do much good. Or that drinking the fermented barley beverage we call beer is going to lower your cholesterol.

What substance is commonly described as eliciting an "umami" taste?

Monosodium glutamate, or MSG, a substance commonly added to many processed foods to improve flavour. This salt of a naturally occurring amino acid has no flavour on its own, but when it is added to foods it elicits a taste that is distinct from the recognized primary tastes of bitter, sour, sweet and salty. Chemistry professor Kikunae Ikeda coined the term umami based on the Japanese word *umai*, meaning "delicious," to describe the savoury sensation imparted to foods by MSG. In the early 1900s Ikeda became interested in why Oriental cooks had the habit of adding seaweed to many dishes. He found that meals prepared in this fashion tasted better, and he succeeded in isolating and identifying the specific chemical responsible for the effect. That chemical turned out to be glutamic acid. It didn't take long for the monosodium salt of this compound to hit the marketplace, and by 1909 Japanese consumers were buying Aji-No-Moto as a white crystalline powder to sprinkle into and liven up their dishes.

At first MSG was produced by breaking down gluten, a wheat protein, into its component amino acids and separating out glutamic acid. Later researchers discovered that certain bacteria were capable of fermenting sugary substances such as molasses into glutamic acid, providing for a much more efficient way of producing the substance.

When MSG is added to foods it is regulated as a food additive, meaning that it has to be shown to be safe in the doses used. Numerous studies have been carried out on the effects of MSG on health without any significant problems cropping up. But as with many additives, or indeed as with many naturally occurring food components, there are people who will have adverse reactions. In this case the migraines, flushes and chest discomfort experienced by these unfortunate souls has been termed Chinese restaurant syndrome because of the propensity of such restaurants to use copious amounts of MSG. There are no lasting consequences, and people who have experienced such effects can patronize restaurants that feature "No MSG Added" signs. Claims on numerous web-sites that MSG is responsible for all ailments in modern society amount to poppycock and leave scientists with a taste that is certainly not umami.

℘

Scientists at the Cocoa Research Institute of Nigeria have developed a chocolate bar that contains about 10 percent starch. Why?

The new chocolate bar doesn't melt till it's about 120°F (50°C), which potentially solves one of the problems chocolate marketers face in tropical countries. Even though the cacao tree is native to the tropics, the finished product is not widely consumed there because chocolate starts to melt at around 75°F (25°C) and becomes a gooey mess. Food scientists at the Cocoa Research Institute found that mixing about 10 percent starch into the choco-late prevents the cocoa butter from melting. And best of all, the chocolate still tastes fine, melts in the mouth and is comparable to ordinary milk chocolate in colour and smoothness.

This is not the first time researchers have tackled the problem of melting chocolate. During the Second World War, the U.S. Army actually commissioned a study to make chocolates melt in the mouth and not in the hand. That challenge was taken up by the Mars and Murray Company of New Jersey, where researcher

Alfred Stern found a way to candy-coat chocolate bits to prevent them from melting. And thus were born M&M's. Although these became very popular, M&M's didn't solve the problem of a non-melting chocolate bar. The issue arose again during the first Iraq War, when the U.S. Army wanted milk chocolate bars that would not melt in the Saudi Arabian desert. Hershey thought it had solved the problem through a secret process, which probably involved incorporating compounds called polyols that retained moisture and raised the melting point. Some 100,000 bars were delivered to the troops but did not meet with great approval. Instead of melting, the bars turned fudgy. They were not reminiscent of the Hershey bars at home. Now it seems that the addition of cornstarch may finally have resulted in a chocolate that melts in the mouth and not in the pocket.

♀

Originally they were called food accessory factors. What do we call them today?

Vitamins. The term "food accessory factor" was coined by British biochemist Frederick Gowland Hopkins in 1906 after he demonstrated that rats fed a diet of proteins, fats, carbohydrates and minerals failed to grow. Although these were known to be the major components of the food supply, they were clearly not sufficient to maintain health. Something was missing. When Hopkins supplemented the diet with minute amounts of milk, the rats thrived. There was something in the milk in addition to the usual nutrients, some "food accessory factor," that was necessary for growth. Hopkins was not the first to make such an observation. Back in

1893 Christiaan Eijkman had discovered that a diet of polished rice caused the terrible disease known as beriberi. It had been known in Southern and Eastern Asia for centuries as a disease associated with weight loss, fatigue and a progressive paralysis of the legs. Death can come from eventual heart failure. Eijkman produced the disease in birds by feeding them a diet of polished rice and then reversed the condition by giving them the rice bran that had been removed in the polishing process. Obviously there was something in the bran that was needed for life, but Eijkman was unable to determine what it was.

Casimir Funk, a Polish biochemist who had come to America, read an article by Eijkman in which he wrote that people eating brown rice were less vulnerable to beriberi than those who ate only the fully milled product. Funk attempted to isolate the substance responsible and finally succeeded in 1912. The compound turned out to belong to a family of molecules called amines, and Funk, thinking these were vital to life, introduced the term *vitamine*. This first vitamine isolated was eventually named thiamine and became known as vitamin B1 when it became apparent that other vitamins were also required for optimal health. Funk suggested that other diseases, like rickets, pellagra and scurvy, were also vitamine deficiency conditions, an idea that had also occurred to Eijkman. As it turned out, not all vitamins belong to the amine family, and eventually the *e* was dropped from the end of *vitamine* to prevent confusion.

In 1929 Hopkins and Eijkman shared the Nobel Prize in Physiology and Medicine for their work on vitamins. Funk, perhaps justifiably, protested that the Nobel committee had given the prize to Hopkins for "his discovery of the growth-stimulating vitamins." Hopkins himself, however, never claimed to be the discoverer of vitamins. Indeed, there was no single discoverer, as many scientists contributed to the knowledge we now have about vitamins. And the book on what these remarkable compounds can do is not yet closed. We now know that they can do more than prevent certain deficiency diseases, and their role as antioxidants and anti-cancer agents is being explored.

☲

A brand of bologna is advertised as being 82 percent fat-free. A 28-gram (1-ounce) slice has 60 calories. What percentage of its calories comes from fat, given that a gram of fat has 9 calories?

The answer is 75 percent. If the bologna is 82 percent fat-free, it is 18 percent fat. Out of 28 grams, then, 5 grams are fat. And 5 grams of fat have 45 calories. So out of a total of 60 calories, 45, or 75 percent, come from fat. This is way more than the 30 percent recommended by health authorities. Similarly, lean hamburger has 60 percent of calories from fat, and lite hot dogs 75 percent. So the 82 percent fat-free claim on the bologna may be technically correct, but it is a lot of baloney!

☲

Why does fried asparagus taste better than boiled asparagus?

The compounds responsible for asparagus flavour are soluble in water but not in oil. If asparagus is boiled, the tasty components leach out into the water, which is normally discarded, but oil does not extract these compounds. Flavour is actually a very complex mix of sensations, encompassing taste, aroma and texture. Raw asparagus contains dozens of compounds which when exposed to heat generate well over a hundred compounds responsible for the distinct aroma and taste of the cooked vegetable.

But perhaps the most interesting bit of asparagus chemistry occurs in the body after the stalks have been consumed and results in the telltale urine fragrance. Apparently this marvel of science wasn't recorded until after asparagus became popular in France, thanks to Louis XIV's penchant for the vegetable. A physician who looked after the royal family noted that "asparagus eaten to excess causes a filthy and disagreeable smell in the urine." And that it does, but not in all asparagus eaters. There is a genetic issue here. The asparagus-urine smell results from a mixture of compounds. Dimethyl sulphide, methyl mercaptan and asparagusic acid are some of the candidates that have been isolated. It seems that whether or not an individual produces these after consuming asparagus depends on how the body metabolizes asparagus compounds, which in turn depends on what enzymes are present, which in turn is controlled by genetics.

To complicate the matter further, the ability to smell these compounds is also genetic, so that some producers may not smell their own production. Apparently 75 percent of the Chinese population cannot smell asparagus compounds in the urine. Taking all studies into account, it seems that roughly 40 percent of the world's population metabolizes asparagus in a fashion that yields compounds in the urine that some but not all others can smell.

Stockton, California, would be an interesting place to research this matter further, it being the asparagus capital of the world. Every year the town hosts an asparagus festival complete with an asparagus-eating contest. The current champion is Joey Chestnut, who downed an amazing 8.6 pounds (4 kilograms) of tempura deep-fried asparagus in ten minutes. Joey is no stranger to competitive eating. In July of 2007, at the famous Nathan's hot dog–eating contest, he knocked off defending champion Takeru Kobayashi by consuming sixty-six hot dogs in twelve minutes.

Competitive eating is most assuredly not a healthy activity, but asparagus, even fried, would have to rank above hot dogs. It contains a significant amount of folic acid, a B vitamin linked to good health. But to satisfy the body's needs, a few stalks will do—8.6 pounds is a bit excessive. I wonder whether Joey Chestnut also holds the record for the world's smelliest urine.

Why is disodium phosphate added to processed cheese?

It acts as an emulsifier and prevents the fat and water from separating, ensuring that the butterfat is evenly distributed throughout the cheese. Processed cheese, also known as American cheese, was invented and patented by James Kraft, the founder of the Kraft food company. Back in the early 1900s, the most popular cheese was Cheddar, but its taste and texture were variable. This is no surprise, since the taste of cheese depends on many factors, including the type of milk, the specific bacterial cultures used to develop flavour and the length of the aging period. Kraft wanted to give his customers a cheese that always tasted the same, had a reliably smooth consistency, melted readily and could withstand elevated temperatures without spoiling. After much experimentation he found that blending various grades of Cheddar cheese and treating the mix with steam yielded a mixture that had most of the properties he was looking for. The high temperature pasteurized the cheese and inactivated any bacteria that would cause further reactions that affect the flavour. One problem, though, was that the finished product was not as smooth as he would have liked, mostly because the fat content was not evenly distributed throughout the cheese. In other words, the cheese wasn't adequately emulsified.

Cheese products always consist of an oil phase, containing fats and oil-soluble compounds, and a water phase, which contains water-soluble proteins and minerals. These phases are incompatible and tend to separate unless emulsifiers intervene. These molecules have both fat-soluble and water-soluble parts and are therefore attracted to both the oil and the water phases, creating an emulsion, which basically is a suspension of tiny droplets of fat in water. Cheese contains casein proteins, which act as natural emulsifiers. One end of these proteins is fat soluble, the other contains calcium phosphate groups, which are water soluble. If the calcium is replaced by sodium, the emulsifying properties improve. Kraft discovered this when he added disodium phosphate to his mix. This substance not only furnished the required sodium, it also provided phosphate, which binds the calcium that has been released from the casein proteins. And that is the secret behind cheese that looks like plastic, keeps like plastic and melts like plastic. Some would even say it tastes like plastic. Interestingly, phosphate here serves the same purpose as it does in laundry products, where it binds calcium and magnesium, which would interfere with the activity of the detergent.

to your health

What is the cause of ouch-ouch disease?

Cadmium poisoning. The name of the disease comes from the painful sounds made by its Japanese victims as cadmium built up in their joints and spines. This classic case of environmental toxicity can be traced back to about 1912, but its cause was not identified until the 1960s. Something was going on in the vicinity of the Jinzu River and its tributaries, that much was clear. People were getting sick, screaming in pain and dying prematurely. But why? Suspicion fell on the river and the mining companies that for years and years had been disgorging their wastes into the water. The mountains upstream were rich in minerals that contained silver, lead, copper and zinc, and mines had been operating there for hundreds of years. As demand for these metals increased in the twentieth century, more and more mining wastes found their way into the river, including increased amounts of cadmium ores. River water was used for irrigating the rice fields, and since rice absorbs cadmium effectively, the metal accumulated in the food supply and consequently in the bodies of the population. The result was ouch-

ouch disease. Although cadmium was identified as the cause only around 1965, by the late 1940s it had become obvious that the disease was linked to the water supply, and mining companies began to store their wastes instead of releasing them into the river. This prevented any more people from contracting cadmium poisoning, but nobody really knows how many people may have been affected since mining operations began to pollute the Jinzu River in the sixteenth century.

In 1929 Dr. Philip Hench noted that one of his elderly arthritic patients improved dramatically after a bout with jaundice. This observation led to the development of the first effective medication against arthritis. What was that medication?

Cortisone. In the 1920s arthritis was a great mystery, with many physicians believing that it was caused by some sort of infection. But in 1929 Philip Hench at the Mayo Clinic in Rochester, Minnesota, made his remarkable observation. A sixty-five-year-old patient who had suffered terribly from arthritis began to improve the day after he came down with a liver problem that caused him to be jaundiced. Hench wondered whether the disease was causing the body to produce some substance that either had an antibacterial effect or was correcting some chemical deficiency. He now began to observe other arthritic patients and noted that they often improved after jaundice as well as after different kinds of surgeries. And he made another critical observation: the fatigue experienced by arthritics was very similar to that suffered by patients with Addison's disease, a condition in which the adrenal glands fail to function properly. Back in the nineteenth century Thomas Addison had noted a relationship between shrivelled adrenals and increased susceptibility to infection and had hypothesized that the adrenals must produce some substance that helps the body deal with stress. Hench now began a search for such a "substance X," as he called it.

Tadeusz Reichstein in Switzerland and Edward Kendall, Hench's colleague at the Mayo Clinic, eventually isolated a number of steroids from adrenal glands, one of which they identified in 1935 as cortisone. This was not easy work; in one case 3,000 pounds (1350 kilograms) of animal glands were required to isolate just 1 gram of a compound. The U.S. government took great interest in this research because of a rumour that Luftwaffe pilots were able to fly at high altitudes thanks to injections of adrenal extract, and U.S. intelligence had reported that Germany was buying large numbers of adrenal glands taken from cattle in Argentina. The Germans probably were doing adrenal research, but the rumour about the Luftwaffe pilots was false.

It was another decade before chemists at the Merck pharmaceutical company were able to synthesize cortisone in sufficient amounts to experiment with therapeutic applications. Finally, in 1948, Drs. Charles Slocomb and Howard Polley at the Mayo Clinic injected a rheumatoid arthritis patient, who had insisted on being a guinea pig for whatever experimental treatment was available, with 100 milligrams of cortisone. The pain relief was almost miraculous. They then tried cortisone in a number of other patients, including one who had been totally bedridden but upon treatment got out of bed and attempted to dance. Another one took seven baths in one day to compensate for the ones she had missed. Unfortunately, not enough cortisone was available to continue treatment, and the patients relapsed. But it was clear that a breakthrough had been achieved.

Today, cortisone and its chemical relatives, referred to as gluco-corticoids, constitute an important class of drugs for the treatment of a variety of illnesses. As the term implies, these drugs increase blood levels of glucose, the substance the body uses as a source of energy, such as that needed to fight off infections. Higher doses of glucocorticoids, however, have the opposite effect and suppress the immune system. Since inflammation is characteristic of increased immune activity, cortisone can serve as an anti-inflammatory agent. That's why it is useful in the treatment of inflammatory conditions such as arthritis, eczema, asthma and Crohn's disease. As always, there is a cost to be paid for the treatment. The increased level of glucose in the blood can lead to type 2 diabetes and depression of the immune system can impair its response to attacks by viruses and bacteria. Weakening of bones, possibly leading to osteoporosis, is another possible side effect, as is the buildup of fatty tissues causing a "moon face."

Philip Hench, Edward Kendall and Tadeusz Reichstein were well deserving of the Nobel Prize for Medicine that they shared in 1950. Hench used some of his prize money to send the nursing supervisor who had handled his arthritis patients to Rome for an audience with the Pope. It is interesting to note that Dr. Hench was a great fan of Arthur Conan Doyle's Sherlock Holmes stories. He demonstrated deductive reasoning much like the great detective in discovering the role of cortisone in the body.

♀

What disease was common to Charlemagne, Henry VIII, Benjamin Franklin and Leonardo da Vinci?

"Patrician malady," or gout. This arthritic condition causes spells of intense pain, most often in the big toe, although other joints can be affected as well. Centuries ago people realized that the disease had a dietary connection. Over-consumption of meat and alcohol seemed to trigger the symptoms. That's why gout was more likely to affect the rich: they were the ones who could afford to be gluttonous.

But today meat eating has spread, and so has gout. It now affects over five million people in North America alone, mostly men and postmenopausal women. Scientists have long known that gout develops when crystals of uric acid build up in joints. Uric acid is produced in the body as a metabolite of purines, compounds found in meat, particularly organ meats such as liver, brains, kidneys and sweetbreads. Seafood is also rich in purines, as are many legumes. The deposition of uric acid in joints causes inflammation of the surrounding tissue, which is why anti-inflammatory drugs such as ibuprofen or naproxen ease the pain of an attack. Colchicine, extracted from the autumn crocus, also has anti-inflammatory effects and has been a traditional treatment for an attack of gout. In severe cases cortisone, sometimes directly injected into a joint, helps. Alcohol consumption, on the other hand, as well as kidney problems worsen gout by slowing down the body's elimination of uric acid.

To pin down the link between gout and dietary purine, researchers followed 47,150 men for twelve years. Men who consumed the most meat, including chicken and organ meats, were 41 percent more likely to develop gout, and each daily serving of fish or shellfish increased the risk by 7 percent. Beer fared the worst, with each daily serving elevating the risk by a whopping 49 percent! Oddly, other alcoholic beverages had a smaller effect. The real surprise came in the form of dairy products and the astonishing protection they offered. The men who consumed low-fat dairy products most frequently had just half the risk of gout when compared with men who ate the fewest such products. Each daily serving of skim milk or low-fat yogurt reduced the risk by about 21 percent. Neither high-fat dairy products nor purine-rich vegetables appeared to alter the risk. This study provides scientific validation for the suspected relationship between gout and meat consumption. Meat-rich modern diets are already implicated in rising global epidemics of type 2 diabetes, obesity and heart disease. It now appears that gout can be added to the list. Meat eaters who hear this news may be tempted to drown their sorrows in beer. But according to this study, skim milk would be a better choice.

What commercial substance can be made by exposing pigskin to light?

Vitamin D. A vitamin is a substance whose presence is crucial to the normal everyday life function of animals but which cannot be directly produced by the animal's body. Accordingly, the daily requirements must be met through regular dietary intake. Vitamin D, however, presents somewhat of an enigma because under the right conditions it *can* be synthesized in the body. There certainly is no enigma about the importance of vitamin D: its presence in the bloodstream is crucial for bone formation. Without vitamin D, calcium cannot be properly absorbed from the digestive tract and cannot be incorporated into bones. We have recently learned that vitamin D performs other important functions in the body as well, possibly reducing the risk of cancer. But vitamin D does not have to be present in the diet if there is sufficient exposure of the skin to sunlight. That's because the skin of many animals—humans among them—has a high concentration of cholesterol, which is converted by enzymes in the skin to 7-dehydrocholesterol.

Exposure of skin to sunlight for regular intervals results in the photochemical conversion of 7-dehydrocholesterol into vitamin D_3. In the absence of sufficient exposure to light, vitamin D needs must be supplied by the diet. Animal products such as saltwater fish and fish liver oils are good sources of vitamin D_3, and small quantities are also found in eggs, veal, beef, butter and vegetable oils. Plants, fruits and nuts are extremely poor sources of vitamin D. To ensure adequate vitamin D intake by the general population, a decision was made in the 1940s to fortify milk with the vitamin. This meant that commercial methods had to be developed to produce the vitamin in large amounts. The key step in the production is to expose 7-dehydrocholesterol to ultraviolet light. Using an organic solvent, this compound can be extracted from the skin of cows, sheep or pigs. Once the 7-dehydrocholesterol has been purified, it is impossible to determine its source, and so religious authorities do not rank it as an animal product. Therefore, although kosher dietary laws forbid the mixing of dairy and meat products, the addition of vitamin D to kosher milk is not an issue.

Why is denatonium benzoate added to some cleaning agents?

To deter children from consuming them. Denatonium benzoate, the most bitter substance known, has a sweet side. It can prevent children and animals from being poisoned. This substance is so bitter that a solution containing just 10 parts per million is undrinkable. And that of course is the idea. Denatonium is an effective deterrent when added to potentially toxic consumer products that children or pets might taste. The number of children and pets poisoned by various household substances is measured in thousands every year, so there certainly is a great deal of commercial interest in preventing such problems.

The history of denatonium can be traced back to 1958, when a Scottish pharmaceutical company was working on developing new versions of local anaesthetics. As is commonly the case, chemists were trying to improve upon existing compounds by slightly altering their molecular structure. In this case, the model compound was lidocaine, a good local anaesthetic. One of the modifications synthesized, eventually to be called denatonium benzoate, was unsuccessful as an anaesthetic, but it had an extremely bitter taste. Testing showed that in spite of its horrible taste, the substance was not toxic.

An idea immediately occurred. Could it not serve as a substance to denature alcohol? Alcohol is a widely used solvent in the chemical industry, but of course is also an important ingredient in beverages. When used as a beverage, alcohol is highly taxed. Such taxes do not apply to industrial alcohol, and this means that there have to be safeguards to ensure that industrial alcohol does not make it into consumers' mouths. One way is to denature it—that is, to add a substance that makes it unconsumable. Denatonium fits the bill. In fact, its name is based upon the term *denatured.* The substance is also added to cleaning agents, pesticides, nail polish remover and antifreeze. It can be used to discourage nail biting. And it can even keep pigs from nibbling on each other's tails. Ethylene glycol, the active ingredient in most antifreezes, is highly toxic but is unfortunately sweet tasting. Pets love to lap up spills, often with tragic results. A teaspoonful can kill a cat, and not much more is needed to kill a large dog. Unfortunately the same goes for children.

However, if denatonium is added, the product becomes so bitter that instead of being swallowed it is immediately spit out. In the United States, pending legislation will force ethylene glycol manufacturers to add it to their products. There is some controversy about this move, because not everyone is convinced that animals are deterred the same way as humans. Since denatonium is put into rat poisons, rats obviously are not bothered by the taste. Still, as far as humans go, there is no doubt that there are children alive today who have been saved by this most bitter of all substances.

♀

For what medical condition would you be advised not to drink tap water without letting it run for a while, especially in the morning?

Wilson's disease. In this rare genetic disorder the body cannot adequately eliminate copper because of a defect in an enzyme that is normally responsible for this job. The full-blown defect shows up only if a faulty gene has been inherited from both parents. If only one parent has the gene, the child is a carrier of the disease and experiences only mild symptoms or none at all. Wilson's disease, however, is not necessarily inherited and can be caused by a spontaneous gene mutation. Whatever the cause, this is a serious disease. Normally, the liver processes excess copper into bile, which is then excreted in the stool. But in Wilson's disease copper ions accumulate and catalyze the formation of hydroxyl free radicals, which damage the liver, nervous system, brain, kidneys and the eyes. Often the first signs, which begin to show in the teenage years, are tremors, depression, anxiety, drooling, loss of appetite and slow, jerky movements. Once the disease is suspected, diagnosis is not difficult. Copper shows up in the urine and in the eyes, where it deposits as brownish rings around the cornea.

Once Wilson's disease is confirmed, foods high in copper, such as chocolate, beans, nuts, whole wheat products and liver, must be avoided. Tap water, which can contain significant amounts of copper that has leached from the copper pipes or brass fittings, is a real risk, especially if it has been sitting in the pipes overnight. In such a case, the maximum copper concentration of 1.3 parts per million as determined by health authorities can easily be exceeded. But avoidance of the intake of copper from food and water is not enough. Excess copper has to be removed from the body by chelating agents such as penicillamine, which bind the metal effectively. Commonly, lifelong zinc supplements are prescribed because zinc prevents the absorption of copper. When treatment is initiated, symptoms resolve in a few weeks. However, if Wilson's disease is not diagnosed, it is fatal. The irony is that copper is an essential nutrient, and we must have tiny amounts in our diet to survive.

One more thing: although the wearing of copper bracelets to counter the pain of arthritis has never been proven scientifically, many people claim to get relief from them. But this is not an option for people with Wilson's disease because enough copper can dissolve in sweat and make its way through the skin to cause problems.

♔

Why are scientists who are experimenting with genetically modified foods interested in "daltons"?

Genetic modification involves the transfer of a gene from one organism to another. Genes are the segments of DNA molecules that instruct cells as to which proteins they should produce. Corn, for example, can be genetically modified with a gene from the *Bacillus thuringiensis* bacterium that codes for the production of an insecticidal protein. "Bt corn" can therefore protect itself against pests and requires less pesticide application. Since many allergens are proteins, any time a gene is transferred there is a risk that a novel allergen can be introduced. For example, when scientists attempted to improve the nutritional qualities of soybeans by transferring a gene from Brazil nuts, it turned out that people allergic to Brazil nuts reacted to the soybeans. The soybeans in question were destined for use only in animal feed, but were tested for allergenicity just in case the beans accidentally entered the human food supply. As a result of the testing, these beans were never marketed.

Critics of genetic modification argue that whereas the allergenic soybeans were caught, other allergens may not be found until it is too late. Actually, scientists have a pretty good handle on what kinds of proteins are candidates for producing allergic reactions and carefully screen genetically modified foods for these. A protein that is not easily broken down when tested under conditions that mimic the human digestive system is a good candidate for allergenicity. Such a protein is more likely to pass through the intestinal wall into the bloodstream, where it can interact with cells of the immune system. Specific amino acid sequences in proteins that cause allergic reactions have also been identified and are catalogued in data banks. Amino acid sequences in novel proteins introduced into foods can then be compared with these. The size of protein molecules, as determined by their molecular weight, is also a clue towards their allergenic potential.

This is where daltons come in. Named after John Dalton, the British scientist who introduced the idea that elements were made up of atoms, a dalton is the fundamental unit of molecular mass. One dalton is defined as one-twelfth the mass of an atom of carbon-12 (the carbon isotope that has six protons and six neutrons in its nucleus) and is equivalent roughly to the mass of a hydrogen atom. Most allergens are known to have molecular weights in the 10,000-to-40,000-dalton range. The bottom line is that proteins introduced by genetic modification are more thoroughly investigated for potential allergenicity than naturally occurring proteins in food. There is no evidence that any human has ever shown any allergic reaction to any novel protein present in any marketed genetically modified food.

❦

There is a theory that the legend of the vampire can be traced to a real human disease. What is that disease?

Porphyria. And it's an unfortunate connection, because people with the disease have to deal with the affliction as well as with the far-fetched vampire connection. The porphyrias are actually a group of inherited diseases with a common feature: the impaired production of heme, which is part of the hemoglobin molecule needed to transport oxygen around the body. In 1985 David Dolphin, a chemistry professor at the University of British Columbia, theorized that people who suffer from porphyria are extremely sensitive to sunlight and risk skin disfigurement upon exposure. So obviously they would only venture out at night. The disfigurement would affect the gums, which recede, making the teeth look like fangs. Actually, only a very rare form of porphyria, congenital erythropoietic porphyria, causes such possible facial changes. Dolphin also theorized that porphyria, being a blood disease, may have stimulated attempts at cure by drinking blood. People with the disease certainly do not thirst for blood. He also suggested that garlic worsens the condition, but there is absolutely no evidence for this. It is tempting to link the origin of legends to real-life events, but the connection between vampires and porphyria is far fetched.

&

What medical term derives from the Latin word meaning "I please"?

Placebo. The placebo effect is one of the most mysterious phenomena in medicine, but it is also one of the most useful ones. Simply stated, a placebo is a substance that pleases the patient by alleviating symptoms in spite of having no physiologically active ingredient. But that does not mean it cannot prompt a real physiological response. After all, the mind can have a powerful effect on the body. People suffering terrible pain will sometimes respond to a sugar solution as dramatically as if they had been given morphine. This is especially so if they have routinely used morphine successfully and then have a morphine dose replaced by placebo. The painkilling effect is the same. The effectiveness of morphine can be countered using a drug called naloxone, which is referred to as an opiate antagonist. For example, if naloxone is administered to a patient whose pain is controlled by morphine, the pain returns. Amazingly, this also happens in patients whose pain is controlled by a placebo. The obvious interpretation is that the placebo triggers an actual physiological effect, perhaps the release of endorphins, the body's own painkilling substances. The effect may originate in the mind, but it is very real.

Placebos can do more than control pain. Experiments have shown that a saline solution can reduce stiffness and muscle tremors in Parkinson's patients and that pills with no active ingredients can slow down the progress of multiple sclerosis. So it should come as no great surprise that over the ages people have found relief in crocodile dung, lizard's blood or oil of earthworms. Neither should we be shocked if today people report that they have experienced wonderful results with juices from the fruits of exotic trees or pills made with special water from the Himalayas. The placebo effect in action.

§

People who have been vaccinated against hepatitis B may have less of a concern about eating peanuts or corn that have been improperly stored. Why?

Improperly stored peanuts, corn and certain other grains can become contaminated with moulds of the species *Aspergillus flavus*. These moulds produce aflatoxins, which are potent carcinogens. The liver is the most likely organ affected, particularly if its function is already compromised by hepatitis B.

Aflatoxins were the first compounds to be recognized as dietary carcinogens. The fact that cancer could be traced to specific substances was first noted in 1761 by John Hill, an English physician, who found that snuff users were more likely to develop nasal tumours. Then in 1775 Percival Pott, an English surgeon, discovered that chimney sweeps were more likely than others to develop skin cancer on their scrotum. It was also noted at the time that chimney sweeps on the Continent did not experience this rare cancer. What was the difference? Bathing was more common on the Continent than in England, and English chimney sweeps were continuously covered in soot from head to toe. Maybe something in soot was responsible for the disease. The gucky material inside chimneys, known as creosote, was the likely candidate. Eventually creosote was confirmed as the culprit, by experiments in which it produced tumours in the skin of animals.

&

What disease results when glutamic acid is replaced by valine in proteins?

Sickle-cell anemia. This is an inherited disease that affects black people almost exclusively. It is characterized by an abnormal form of hemoglobin, the molecule found in red blood cells responsible for transporting oxygen. Sickle-cell hemoglobin cannot carry enough oxygen and therefore causes anemia. But that's not all. The abnormal proteins cause the normally round red blood cells to become crescent, or sickle, shaped. These sickled red blood cells cannot pass through the small blood vessels in the spleen, kidneys, brain and bones and block blood flow, potentially causing organ damage.

Why does this happen? Sickle-cell disease is a molecular disease, an expression coined by Linus Pauling, who carried out some of the early work in this area. The disease is caused by a defect in the hemoglobin molecules. Hemoglobin is very large molecule and consists of four polypeptide chains. That's just the chemical expression used to describe a chain of amino acids. For hemoglobin to function properly, the right amino acids must be attached together in exactly the right way. About 0.3 percent of blacks inherit a faulty gene from both parents resulting in their DNA giving faulty instructions about the synthesis of hemoglobin. Instead of incorporating glutamic acid in a specific position in one of the polypeptide chains, a molecule of valine is inserted instead.

Now this may not sound like much of a problem when we are talking about one wrong amino acid in a chain consisting of 146 amino acids, but it is. The valine molecules form strong attractions to other valines in nearby hemoglobin molecules, and this causes the hemoglobin polypeptide chains to aggregate and precipitate out of solution. This is what then alters the shape of the red blood cells. Any activity that increases the need for oxygen can make the situation worse since the abnormal red blood cells cannot meet the oxygen demand. Vigorous exercise, mountain climbing or flying at high altitudes with insufficient oxygen fall into this category. The lack of oxygen transport and the clogging of blood vessels by sickle cells can result in terrible abdominal pain and shortness of breath.

The spleen is one of the first organs to be damaged by sickle-cell disease, and since it is an important part of the immune system, sickle-cell patients develop infections more readily. A sickle-cell crisis requires treatment with lots of fluids and pain-relieving drugs. Sometimes blood transfusions are needed as well as extra inhaled oxygen. There is no effective drug to treat sickle-cell anemia, but there have been some encouraging results with urea, which can get between the troublesome valine molecules and prevent the polypeptide chains from aggregating. A molecular cure for a molecular disease may be in the offing.

☺

A newly diagnosed cancer patient scoured the Internet for information about what to do in addition to following her physician's advice about standard therapies. She came to the conclusion that supplementing her diet with large doses of lima beans was a good idea. Was she right?

No. These days it is common for anyone who has been diagnosed with an ailment to Google it. That is exactly what the woman in question did, and she soon found repeated references to the use of a substance called laetrile in the treatment of cancer. The argument is that laetrile, sometimes nonsensically called vitamin B-17, releases cyanide in the body, which in turn destroys cancer cells. It rose to fame in the 1950s, when proponents claimed it could cure cancer. Unfortunately, animal studies failed to find any beneficial effect, but the public, understandably hungry for an effective cancer treatment, put the pressure on the government to mount human studies.

In response to the political pressure, the National Cancer Institute sponsored a number of trials on patients with advanced cancer for whom no treatments had worked. Laetrile was found to be useless.

Its proponents claimed that the substance had not worked because these patients' immune system had already been destroyed by radiation or chemotherapy. Because of the lack of evidence for any benefit, laetrile was eventually made illegal. That has not stopped it from being sold on the Internet or from being used by clinics in Mexico.

Laetrile is a synthetic analogue of amygdalin, a compound that occurs naturally in apple seeds, apricot pits and lima beans. It breaks down to release hydrogen cyanide. This is the information that prompted our cancer patient to start eating lima beans, hoping the cyanide they released would destroy her tumour. There is no possibility of this happening, but there is a possibility of getting sick from eating large amounts of lima beans, especially if they are consumed raw. Enough cyanide can be released to cause serious nausea, vomiting and dizziness. If the beans are cooked, the risk is greatly reduced, because hydrogen cyanide dissipates into the air. Even though there is absolutely no evidence that laetrile, or any other source of cyanide, is beneficial in the treatment of cancer, its proponents have not given up and fill websites with misleading information .

℘

Alka-Seltzer is often taken by people to relieve excess stomach acidity, yet the product contains citric acid. What is the role of the citric acid?

Alka-Seltzer contains aspirin, sodium bicarbonate and citric acid. The role of citric acid is to react with the bicarbonate to produce the fizz that impresses people. It does little else. Sodium bicarbonate is better known as baking soda because it can cause baked goods to rise as it liberates carbon dioxide gas upon reaction with an acid. In Alka-Seltzer some of the baking soda reacts with citric acid to produce the fizz, but since there is an excess of bicarbonate, enough is left over to neutralize excess stomach acidity. The aspirin of course is effective against headaches, so that Alka-Seltzer is a useful medication for someone simultaneously suffering from an upset stomach and a headache, as can happen after a particularly pleasurable night out. To witness the efficiency of gas production, just take an Alka-Seltzer tablet, place it in a film canister with a little water, put the lid on and step back. You'll see the power of carbon dioxide as the lid is blown off in a spectacular fashion.

♀

Why are researchers at Sun Yat-sen University in China feeding "forbidden" rice to mice and rabbits?

Forbidden rice derives its name from the fact that in ancient China it was forbidden to everyone except the Emperor. That's why it is sometimes also called Emperor's rice. It looks very different from ordinary rice: it is black. It was reserved for the Emperor because of a belief that the black rice had properties that enhanced longevity. Supposedly it also acted as an aphrodisiac. Why the general Chinese population was forbidden to improve their chances at longevity or at sexual satisfaction is not clear. What is becoming clear, though, is that black rice really does have some health benefits.

The natural black pigment responsible for the colour of the rice belongs to a family of compounds called anthocyanins. These have antioxidant properties and have been linked with all sorts of health benefits, including a reduced risk of clogging the coronary arteries. That is what Dr. Wenhua Ling was interested in investigating. He first fed his mice and rabbits a cholesterol-rich diet to promote heart disease. Then half the animals received a supplement made from white rice bran while the other half was treated with black rice bran. Animals in the black rice bran group had fewer oxidation products in their blood (a good thing) and experienced less artery-clogging plaque. Experiments are under way to see how humans benefit from black rice. Maybe we should change the name from "forbidden" rice to "recommended" rice. Bring it on! It's more expensive than regular rice, but probably worth it.

♀

What gem can be used to strengthen bones?

The pearl. Pearls are made of calcium carbonate, the active ingredient in many calcium supplements. Of course, it is a lot cheaper to swallow a Tums tablet than to eat your pearl earrings. But as far as the body goes, there is no difference. The shells of oysters, which harbour pearls, are also made of calcium carbonate, and many commercial calcium supplements advertise that they are made of "natural" oyster shells. This is supposed to entice people who think that natural substances are superior to synthetics. Actually, in some cases oyster-shell supplements have been found to be contaminated with lead. Calcium supplements, no matter what the source, help build bones. But traditional Chinese medicine maintains that pearls can do more than that. Pearl powder, according to traditional doctors, can be used to treat epilepsy, convulsions, hyperactivity, hypertension, insomnia and palpitations. No controlled studies corroborate such use. Neither is there evidence that a pearl face mask can rejuvenate dry, dull skin. Apparently believers in this beauty treatment mix a teaspoon of pearl powder with enough olive oil to form a paste and apply it to the skin for ten minutes. They could probably improve their appearance more by wearing a pearl necklace.

♀

Cattle dropping dead on Canadian farms in the 1920s led to the discovery of which widely used drug?

Warfarin, a commonly used anticoagulant. It is perhaps best recognized by one of its trade names, Coumadin. Often erroneously called a blood thinner, warfarin prevents the formation of blood clots by blocking the production of vitamin K by the liver. A number of medical conditions can cause the formation of blood clots, which in turn can have catastrophic consequences. People who have had warning strokes (transient ischemic attacks, or TIAs) are commonly put on warfarin, as are patients who have undergone heart valve replacements or who suffer from irregular heartbeats.

In the early 1900s Canadian farmers began to grow sweet clover as "green manure" to improve soil quality, and sometimes ended up using the crop as hay for cattle feed. Unfortunately, sweet clover hay sometimes killed cattle.

A Canadian veterinarian, Dr. Frank Schofield, eventually unravelled the mystery. The problem arose only when the hay became contaminated with a mould. When this occurred, a compound in clover called coumarin was converted to dicoumarol, which had a very significant anticoagulant effect. Affected cattle hemorrhaged internally and died. Karl Link at the University of Wisconsin became interested in dicoumarol, successfully synthesized it in the lab and patented it as a drug. Eventually it was commercialized as warfarin . But warfarin had an application even before it was ever used on human patients. Seeing how quickly it killed animals, Link had an idea. Why not use dicoumarol as a rat poison? The idea turned out to be a good one. Warfarin became a huge success and is still commonly used today to do away with annoying rodents. To pursue his animal studies, Dr. Link had received a research grant from the Wisconsin Alumni Research Fund, which went under the acronym WARF. He was so thankful for this support that he named the commercial product in WARF's honour.

♀

One of the most bizarre medical conditions is aquagenic pruritus. What is it?

Itching in response to exposure to water, usually a shower. Aquagenic means "arising from water" and pruritus means "itch." Luckily this condition is very rare, but when it strikes it is debilitating. Sufferers describe the prickling discomfort that lasts on average for forty minutes after being triggered by water exposure as virtually unbearable. Numerous treatments have been tried, but no cure has been found. Symptoms can often be alleviated with a cream based on hot pepper extract. The active ingredient, capsaicin, dulls sensations. Antihistamines can also help, as can treatment with ultraviolet light. Some patients respond well to a sodium bicarbonate bath, which is interesting given that the symptoms are brought on by water. Even drugs such as atropine that block the action of acetylcholine, an important neurotransmitter in the body, have been tried, with some degree of success. Given that opiates, such as morphine and heroin, sometimes cause prickling sensations, physicians have experimented with drugs that block opiate activity, on the off chance that they may help relieve aquagenic pruritus. It turns out this works. Naltrexone, which is an opiate antagonist, was found to suppress the sensation of itching, including that caused by water.

So now there is some hope for the unfortunate few for whom showering is a torturous experience.

⚕

What are pumpkin seeds used for in medicine?

To improve urinary flow in men affected by prostate problems. Clinical studies have shown that pumpkin seeds can help when the urinary tract becomes partially obstructed by an enlarged prostate or by prostate cancer. The seeds do not cure the underlying problem, they just help with the symptoms. About 10 grams a day will do the job, which means 1 to 2 heaping teaspoonfuls (5 to 10 millilitres) of ground pumpkin seed with liquid in the morning and evening. Exactly what component in pumpkin seeds is responsible for the benefit is unknown, but phytosterols are likely candidates. Maybe Charlie Brown knew more about the Great Pumpkin than he let on.

⚕

What is the common name for the plant-derived medication that has classically been used to induce vomiting in cases of poisoning?

Syrup of ipecac. This extract of the roots of the South American ipecacuanha plant has traditionally been stored in medicine cabinets and in emergency rooms to induce vomiting in case of poisoning.

The wisdom of using ipecac has been questioned because activated carbon usually is a safer and more effective treatment. In any case, a poison control centre should be consulted before ever using syrup of ipecac since it certainly is not appropriate for all poisons. Corrosive substances such as acids or bases, paint thinner, gasoline or kerosene can cause severe damage when vomited. Determining the dose of ipecac needed is also difficult, and overdose can cause heart problems.

Ipecac has been used to treat amoebic dysentery, a usually tropical gastrointestinal disease caused by an amoeba, a tiny parasitic creature. The main symptom is bloody diarrhea. Interestingly, Brazilian Indians have long been aware of this use of ipecac and brought it to the attention of Portuguese missionaries, who then introduced the treatment to Europe. The active ingredient, both for inducing vomiting and for killing amoebas, is emetine. Using the right amount of ipecac is critical, and one wonders how many people were killed by the syrup relative to the number who were cured of dysentery. Ipecac is no longer used for dysentery and is rarely used for poisoning.

Unfortunately, ipecac is sometimes used by anorexics and bulimics to lose weight, a dangerous practice.

♀

Until the early part of the twentieth century, the American Midwest was known as the "goitre belt." Why?

There was a lack of iodine in the soil, so crops lacked the iodine needed for proper functioning of the thyroid gland. The thyroid gland, located in our neck, produces hormones that control the rate of energy production in all cells in our body and therefore influence the functioning of all organs. Insufficient production of thyroid hormones can cause weight gain, lethargy, constipation, cold clammy skin and perhaps even hair loss and premature greying. In utero it can cause cretinism.

There are various reasons for hypothyroidism. The most common form is known as Hashimoto's disease and is caused by chronic inflammation of the thyroid gland. It is treated by giving the patient thyroid hormone. But there is another possible cause for hypothyroidism: lack of iodine in the diet. Goiter – a swelling of the thyroid – used to be endemic in the American Midwest, while the condition was almost never found near coastal areas. By 1920, scientists had established that the connection involved iodine. Because of the lack of iodine in the soil, crops, and the animals that fed on them, were deficient in iodine. This also explained why goitre was seen in areas where cabbage was a staple in the diet. This vegetable contains isocyanates, which interfere with the absorption of iodine.

When people found out about the iodine connection, they decided to take matters into their own hands. Or at least to hang them around their necks. They began to wear bottles of iodine. Some started to eat iodine to the extent that they produced a *hyper-*thyroid condition. And then in 1924 Michigan began to experiment with adding sodium iodide to salt, and Rochester, N.Y., added it to drinking water. The solution to the goitre problem was under way.

Today, salt with added potassium iodide—called iodized salt—is common. But there is a problem. Iodide is slowly converted to iodine in moist air. Since iodine is volatile, salt slowly would lose its power to protect against goitre. It is therefore common practice to add iodine stabilizers to iodized salt. These are substances that convert iodine back to iodide. Sodium thiosulphate used to be added, but it sounded too chemical and people became worried. Manufacturers switched to dextrose, which is as effective and sounds more innocuous. Sodium bicarbonate is also added because the oxidation of iodide occurs readily in an acid solution and not in a base. The bicarbonate produces basic conditions. Sometimes disodium phosphate or sodium pyrophosphate are used to provide the alkaline conditions. These are also sequestering agents, which bind trace metals that catalyze the oxidation of iodide to iodine.

Unfortunately, this easy protection is not being carried out everywhere. In India an estimated 250 million people suffer from iodine deficiency, which results in decreased motor skills, low IQ and poor energy levels. About 9 million Indian children a year are born as cretins. Because of the humid air, salt is usually sold in large crystals, which resist humidity. These are sometimes sprayed with potassium iodate, but it makes the crystals look dirty, and so people wash the salt before crushing it.

♀

We may learn more about Parkinson's disease from the experience of a former stuntman with early-onset Parkinson's who found relief after taking an illegal substance. What was it?

Methylenedioxymethamphetamine, better known as ecstasy. Former British stuntman Tim Lawrence has early-onset Parkinson's disease. He had been taking the classic medication L-dopa, which helped unfreeze his immobile body but gave him wild, flailing movements called dyskinesias. Parkinson's disease is characterized by a deficiency of the neurotransmitter dopamine in the brain, a deficiency that can be temporarily remedied with L-dopa. The efficacy of this drug, however, is lost with time. One day Lawrence went to a night club with friends and sampled some ecstasy. To his surprise, and delight, his disease seemed to magically vanish, at least for a brief time. He experimented with the drug again and found that within two hours he was able to perform gymnastics, something he had been unable to do for years! Unfortunately, ecstasy is a dangerous drug capable of causing sudden potentially fatal rises in body temperature. But Lawrence's experience has stimulated novel lines of Parkinson's research. It seems that there is more involved in the disease than just a deficiency of dopamine. Perhaps coupling L-dopa with some type of amphetamine will be better at easing the symptoms of Parkinson's than L-dopa alone. But nobody is suggesting that Parkinson's patients self-medicate with ecstasy.

We know that Popeye is energized by eating spinach. It probably also reduces his risk for macular degeneration. Why?

Green leafy vegetables like spinach are rich in lutein and zeaxanthin, compounds that can prevent light from damaging the macula, the central part of the retina responsible for straight-ahead sight. Degeneration of the macula is the leading cause of blindness in people over the age of sixty-five. People who frequently consume carotenoids, a family of compounds found in a variety of fruits and vegetables, have a reduced risk of developing the condition. Carotenoids are named after beta-carotene, found in carrots, but in the case of preventing macular degeneration two other members of the family, lutein and zeaxanthin, are more effective. According to the most likely explanation, these compounds form the the yellow pigment in the retina that filters out damaging blue light. *Luteus* is Latin for "yellow," so we shouldn't be surprised that lutein is found in egg yolks, corn, squash, pumpkin and orange peppers. One study in which 356 patients with macular degeneration were matched against 520 controls found that people who consumed carotenoid-containing foods at least five times a week had a 43 percent lower risk of macular degeneration.

The juice of the aloe vera plant has a reputation for helping to heal skin disorders. But what did Aristotle recommend the juice for?

Cleansing the body . . . internally. You know what Aristotle said to Alexander the Great in 325 BC? Come here, Alex, I have something to show you. And he showed him a plant. It was the Socotrine aloe, which originated from the island of Socotra, east of the Horn of Africa. Aristotle had learned that the juice of the plant had an amazing effect. It was a purgative. You drank a little and everything came out. This was an important finding at the time because it was widely believed that illnesses could be cured by cleaning out the body. Alexander thought so much of this effect that he sent investigators to Socotra to find out if a purgative could be mass-produced from the plants. Socotrine aloe is just one of about two hundred species of aloes but is particularly potent as a purgative. Other aloes have less of a dramatic effect. But they all have some potential to act as purgatives.

The aloe that is familiar to us is the Barbados aloe, which originated in the Mediterranean regions and was taken by the Dutch to the West Indies for cultivation. It was mostly a decorative plant and sometimes used as a purgative. But it also found use as a remedy for mild skin disorders, burns, scratches and cuts. This effect had actually been described by the Egyptians as early as 1500 BC. Aloe is mentioned in the famous Ebers papyrus, which catalogued drug use at the time. Cleopatra is said to have used aloe to beautify her skin. Today, aloe shows up in a myriad products, ranging from creams to shampoos. Aloe juice is widely promoted on the Internet as an internal cleanser, and some sites even suggest that it can boost the immune system and fight off cancer. Nature's gift to humanity, they call aloe. But I'm always suspicious of webmasters bearing gifts.

The thin, clear gel that leaks from an aloe leaf when it is cut does have some healing properties. Nobody knows exactly which chemicals are responsible for this effect, but it is probably a slew of them. An enzyme called bradykininase destroys bradykinin, a chemical found in our body that produces pain. Magnesium lactate has an anti-itching effect. Other components have antifungal and antibacterial effects. Several studies have shown that aloe gel helps with frostbite, sunburns and wound healing. In test tube studies, fluid from fresh leaves promotes the growth of normal human cells. Whether this effect is retained in processed products that use aloe is more questionable. Some of these products may not have very much aloe to start with—just enough to allow its presence to be declared on the label. But some others have been shown to improve wound healing after facial dermabrasion. The recommendations for internal use are more iffy. The juice is sometimes contaminated with the aloe components that act as laxatives, and this may not be a desired effect. The benefits claimed for healing ulcers and soothing digestive problems have not been confirmed by scientific studies; neither have the

very tenuous claims about cancer treatment. At this point it is better to limit use of aloe to external mild skin problems. I say mild
because studies have shown that for serious cuts, such as after a
Caesarean section, applying aloe may actually retard healing. But for
your everyday burns, abrasions and cuts, fresh aloe gel squeezed
from a leaf may be just what the doctor ordered. So it's a good idea
to keep an aloe plant at home. Even if the skin-healing results are
not that great, you'll have a very pretty plant.

♀

Why did a hospital in Birmingham, England, replace door handles, bathroom taps and toilet-flush handles with ones made of brass?

To try to cut down on the spread of methicillin-resistant *Staphy-
lococcus aureus* (MRSA) bacteria. Hospitals are plagued by the spread
of bacteria that are resistant to antibiotics in spite of rigorous
cleaning efforts and the isolation of infected patients. Researchers
at the University of Southampton have, however, discovered a non-
invasive method that may help cut down on hospital-acquired
infections. Laboratory tests revealed that copper and its alloys have
antimicrobial properties, effectively killing germs that land on the
surface. On stainless steel surfaces, MRSA bacteria remain active
for days, but on brass, an alloy of copper and zinc, bacteria die
within five hours, and on pure copper, they meet their end in thirty
minutes. Preliminary trials show that copper is also effective against
Clostridium difficile bacteria, another hospital scourge. Even flu
viruses are sensitive to copper. If the experiment in Birmingham,
which will include issuing brass pens to doctors, manages to cut

down on infections, doctors and patients will be delighted. So will the copper industry.

Actually, the use of copper in medicine is not new. The ancient Egyptians cooked up copper with vinegar to produce copper acetate, better known as verdigris, which they used to treat eye infections. Pliny, the Roman "natural philosopher," recommended copper oxide to purge the stomach and to kill intestinal worms. It probably did that, but a side effect was that it could also kill the patient. Copper compounds in the wrong dose can be very toxic. But even this can be beneficial, as in the use of copper sulphate, a component of Bordeaux mixture, used to control fungi on grape vines. And then of course there is the story of copper bracelets helping with arthritis. Apparently this can be traced back to an observation in 1939 by German physician Werner Hangarter that copper miners in Finland suffered less from arthritis than the general population. Hangarter went on to test copper salts of aspirin as a treatment for arthritis, but this did not pan out, probably because of copper's toxicity. But the experiment seems to have spawned a whole industry, that of producing copper bracelets to battle arthritis. There is no evidence that these bracelets produce anything other than a placebo effect. But at least you can be sure that they will not transmit MRSA bacteria.

§

What does quaternium-15 do?

It kills micro-organisms. Although "causes contact allergy" would also be an acceptable answer. Quaternium-15 is widely used in cosmetic products, ranging from mascara and moisturizer to shampoos and body washes. Such products can be contaminated by a variety of bacteria, yeasts and moulds, either during processing or by consumer handling. One of the most effective antimicrobial compounds is formaldehyde and is used in a variety of cosmetic products. But there are some problems with formaldehyde. It can dissipate from a product before the product is all used up, and it can also cause allergic contact dermatitis. That's why formaldehyde-releasing substances such as quaternium-15 have been developed. Quaternium-15, used in more than six hundred products, releases formaldehyde slowly and maintains a steady concentration in the product. But it doesn't solve the allergenicity problem. In fact quaternium-15 is the most allergenic of the preservatives used, capable of causing a rash. The allergic reaction can be due to the formaldehyde released or the quaternium-15 itself.

Of course it may not be easy to track down the source of an allergy, since many cosmetic ingredients can cause such reactions. Fragrances, which encompass over a thousand chemicals in common use, are the most likely candidates, but preservatives such as parabens and quaternium-15 are a close second. Parabens are almost ideal preservatives because they are colourless, odourless, and rather non-irritating, and have activity against a wide range of bacteria. They are less active against fungi and therefore are usually combined with another type of preservative, such as a formaldehyde release. The active ingredients in cosmetics can also cause allergic reactions. Cocamidopropyl betaine (CAPB), for example, is a surfactant, a substance that allows oil and water to mix. It is widely used in baby shampoos because it doesn't irritate the eyes and is also found in some creams to prevent oil and water from separating. It also serves as an anti-static agent in hair conditioners. And in some people, it causes allergic contact dermatitis such as scaling around the eyelids after using baby shampoo. The only way to identify the culprit in an allergic reaction is by patch testing.

🍄

The seeds of a plant that were used as the principal ingredient in the original version of Lydia Pinkham's Vegetable Compound are now being explored for their potential to control blood sugar levels in diabetics. What is the plant?

Fenugreek, a small bean-like plant, the leaves of which can be used as a herb and the seeds as a spice. It has a long and fascinating history. Seeds of fenugreek were found in King Tutankhamun's tomb—supposedly they would help sustain him in the other world—and traditional Chinese medicine and Indian Ayurvedic medicine used fenugreek to induce labour and lactation, to aid digestion and as a general wellness tonic. In the nineteenth century Lydia Pinkham used fenugreek seeds in her magical cure-all for female problems.

There is no magic in fenugreek, but the plant may indeed have some legitimate medical applications. The most interesting possibility lies in the control of blood sugar.

Stimulated by anecdotes of diabetes control with fenugreek in India, researchers have begun to investigate the potential therapeutic use of the plant. Animal trials have clearly shown that fenugreek can lower blood sugar levels, and the few human trials that have been carried out lend support to the proposed hypoglycemic (blood-sugar-lowering) effect of the seeds of the plant. Both the powdered seeds and water-alcohol extracts of the seeds have been investigated. A placebo-controlled trial in which type I diabetics were given 50 grams of defatted fenugreek seed powder twice daily did demonstrate a blood-glucose-lowering effect, but the trial involved only ten patients. A larger study of sixty patients with type 2 diabetes used 25 grams of powdered seed twice daily showed a significant decline in fasting blood glucose as well as in hemoglobin AIc, a measure of blood glucose control over several months. A small trial using I gram of a seed extract per day also showed improved glucose control.

Why fenugreek should have any effect on blood glucose isn't clear. As with any plant product, fenugreek seeds contain dozens of compounds, among which are steroids such as yamogenin and diosgenin. These are chemically related to the body's natural sex hormones and may explain fenugreek's traditional use to induce labour and help with female complaints.

Also contained in the seeds is 4-hydroxyisoleucine, which may stimulate insulin secretion. But in addition, fenugreek contains a small amount of coumarin, which may interfere with anticoagulant medications.

All in all, there is some evidence that powdered fenugreek seeds, in the ballpark of 25 to 100 grams a day, may help with the control of blood glucose, but the evidence is not compelling.

Still, there is no doubt that fenugreek can add some flavour notes to a curry. And believe it or not, to artificial maple syrup. Roasted fenugreek seeds develop a maple syrup taste. But this would not be an appropriate way for people attempting to control their blood sugar to experiment with fenugreek.

What medical condition requires a reduction of phosphorus in the diet?

Kidney disease. Phosphorus is an essential mineral required in the diet. It is needed to form calcium phosphate, the major component of bones, and is also an integral part of DNA and RNA, the body's genetic substances, as well as of the main energy-storing compound, ATP. The source of all the phosphorus needed for the synthesis of these essential compounds is phosphorus in the bloodstream, which in turn is supplied by phosphorus in the diet. In the blood, phosphorus is present mostly in the form of phosphate, which is kept in balance with calcium by the kidneys. If phosphate levels rise, the kidneys excrete the excess. But if the kidneys are not functioning normally, phosphate levels in the blood become elevated, and calcium is then withdrawn from bones to keep stride with the increasing amounts of phosphate. This eventually leads to painful and brittle bones.

Kidney patients are therefore advised to follow a low-phosphorus diet and are often prescribed phosphate binders to be taken with meals. The most common one is calcium carbonate, which is also widely used to reduce excess stomach acidity in the form of Tums. As far as diet goes, dairy products, colas, organ meats, legumes, chocolate and whole grains are high in phosphorus and should be limited. Kidney beans, for example, may resemble the kidneys in shape but have to be limited in the diet of kidney patients. And cooking doesn't solve the problem. About 90 percent of the phosphorus is retained after cooking. But kidney patients still have plenty of choices. Potatoes, squash, cabbage, carrots, soy milk and small servings of poultry and fish are fine.

toxic relationships

What chemical created a great deal of controversy when it was found to leach out of clear plastic water bottles?

Bisphenol A. People today can often be seen squinting at the little recycling logo on the bottom of their plastic beverage and food containers looking for the number 7 with the letters PC underneath it. Its presence can raise the alarm and send the consumer scurrying for other containers adorned with the numbers 1, 2, 4 or 5. Why? Because much has been said and written about the potential risks of small amounts of bisphenol A in our environment. Nobody argues that we are not exposed to traces of bisphenol A. In fact, it would be surprising if we were not. It is one of the components used to make polycarbonate plastics, which are ubiquitous. Bottles, compact discs, sports helmets, MP3 players, computers, eyeglasses, electric meter covers and dozens of other items make use of polycarbonate's extreme toughness. Epoxy resins, including the ones that line the inside of some food cans, also are made with bisphenol A. Although bisphenol A is combined with other chemicals to make the final product, it is inevitable that trace residues of this substance are present and can, under certain conditions, leach out.

Of course the question is how much leaches out and whether such amounts present a danger. Some researchers suggest that there is a risk, based on studies with rodents exposed to amounts in the range that people may be exposed to. They claim that bisphenol A mimics the action of estrogen and that it may play a role in conditions ranging from obesity, Type 2 diabetes, impaired brain development, and early onset of puberty, to breast cancer, prostate cancer, miscarriage, sperm defects and impaired female reproductive development. These doses are well below the levels that regulatory authorities deem are safe, to which those concerned about bisphenol A respond that this chemical exhibits different effects at tiny doses than it does at larger ones. They imply that the accepted guiding principle of toxicology, namely that the dose makes the poison, doesn't apply in the case of chemicals that have hormone-like effects and that the dose-response curve, which toxicology predicts should be linear, actually is U-shaped. This is an interesting possibility, and the argument did trigger a number of studies that attempted to repeat the surprising low-dose effects that had been found.

A number of well-designed large-scale studies have failed to confirm the low-dose effect. So we are left with the battling scientists. Each group accuses the other of conducting studies improperly, or at least of using unrealistic conditions. Allegations of vested interests also fly back and forth. There are accusations that untested chemicals are released onto the market and that governments are failing in their duty to protect consumers from toxic substances. Numbers are flung about with great abandon. We are told that some eighty thousand chemicals are registered for consumer use, and that seven new ones are added every day, often without adequate tests for safety. The public is left confused.

Is it possible that bisphenol A has a harmful effect on people? Maybe. But we are exposed to dozens of estrogenic compounds, both synthetic and naturally occurring, every day. Estrogenic compounds from birth control pills, hormone treatments and foods such as soybeans are present in our environment in doses far greater than those of bisphenol A. These are virtually impossible to avoid, but people who are concerned about bisphenol A can in theory reduce their exposure by avoiding beverages stored in polycarbonate bottles and staying away from canned foods. Where this makes the most sense is in the case of babies, who are the most sensitive to chemical insults. Since glass baby bottles are readily available, there is no need to use polycarbonate. But I don't think the hiker who takes along a polycarbonate bottle to quench his thirst has anything to fear from bisphenol A.

♀

What sort of pills can people obtain for free from the government in some American states if they live within 10 miles (16 kilometres) of a nuclear reactor?

Potassium iodide. An accident at a nuclear reactor can release radioactive substances into the environment. Radioactive iodide is of particular concern because it can be inhaled or ingested from contaminated vegetation, dairy products or meat. Iodide concentrates in the thyroid gland, and the radioactive variety emits beta particles and gamma rays that can damage surrounding tissue and cause cancer. If there is a concern about exposure to radioactive iodide, regular potassium iodide can be orally administered. It will be absorbed within an hour and concentrate in the thyroid gland, preventing any further uptake of iodide by the gland. Taken within twelve hours before exposure, potassium iodide can almost completely prevent radioactive iodide from entering the thyroid. There is some protection also if it is taken within twenty-four hours after exposure, but not if more than twenty-four hours have passed. An adult requires about 130 milligrams a day, and children need about half that much.

The dangers of radioactive iodide exposure, particularly in children, were dramatically demonstrated after the Chernobyl accident in 1986. Within four years there was close to a hundredfold increase in thyroid cancer in children in the areas covered by the radioactive plume. Poland, where immediately after the accident potassium iodide tablets were distributed to some eleven million children and seven million adults, serves as a remarkable contrast to the Ukraine situation. Virtually no increase in thyroid cancer was observed, clearly demonstrating the protective effect of potassium iodide.

☙

Some garden hoses are labelled as "not for drinking." Why?

Because they contain lead, which can leach out. Many garden hoses are made of polyvinylchloride, or PVC, which contains small amounts of lead added as a stabilizer. Otherwise these hoses would degrade quickly when exposed to the sun. If water stands in these hoses overnight, the first draw can contain levels of lead anywhere from ten to a hundred times higher than what is allowed in drinking water. There is concern about lead because even tiny amounts can cause health problems. But not this tiny. Drinking occasionally from a hose, even one that contains lead, is not going to cause problems. The warnings seem to be an overreaction. I'm not suggesting people should make a habit of drinking from a garden hose. But I would worry more about the dirt that may have got into it than the lead that might be coming out.

In the sixteenth century enemies of the sculptor Benvenuto Cellini attempted to murder him by poisoning his food. Instead, they ended up curing him of syphilis. What had they put in his food?

Mercuric chloride, also known as "corrosive sublimate." People have been intrigued by mercury and its compounds for over 3,500 years. Ko Hung, an ancient Chinese alchemist, was one of the first to describe how heating red cinnabar, or mercury sulphide, yielded silvery mercury metal. From then on alchemists believed that mercury was the key to transmuting elements into gold, and experimented with all sorts of concoctions to try to achieve this. They of course never made gold but were successful in converting mercury into various mercury compounds. Heating the metal with salt and alum in a closed earthenware pot resulted in mercuric chloride. This was eventually tried in the treatment of syphilis, basically because anything available was tried. By the fifteenth century syphilis had become a huge curse in Europe, and some success was seen with mercuric chloride treatments. Mercuric chloride killed the *Treponema pallidum* bacterium, but unfortunately it usually killed the patient too.

Benvenuto Cellini was an Italian sculptor noted for being the first to produce life-size works in bronze. Whether because of jealousy or some other reason, his enemies conspired to spike his food with corrosive sublimate. Cellini had struggled with syphilis since the age of twenty-nine and had refused treatment with mercury, probably aware of the possible fatal effects. In any case, after eating the poisoned food he became very ill with gastrointestinal problems, but recovered. And when he recovered, the syphilis was gone. Luckily for Cellini the poisoners had not put enough mercuric chloride into his meal to kill him, but had put enough to kill the bacterium responsible for the disease. An excellent example of how only the dose makes the poison . . . or the cure. Mercury compounds are no longer used in medicine, and syphilis readily responds to antibiotics such as penicillin. At least until the bugs develop resistance.

❧

In the nineteenth century, poisoning was a common method of murder. Suspected victims were sometimes autopsied in an attempt to find signs of poisoning. What poison was suspected if yellow patches were found on the stomach?

Arsenic. The yellow patches were arsenic sulphide, which forms when proteins in the body decompose to yield hydrogen sulphide, which in turns reacts with arsenic to form the telltale arsenic sulphide. Historically arsenic trioxide has been the most widely used poison to do away with people because it was readily available as a pesticide and rodent killer. It also had the advantage of producing symptoms that could be mistaken for cholera or dysentery, which were common ailments. A single large dose of arsenic—and 65 milligrams or so would be considered a large dose—causes symptoms rapidly. Burning in the throat and terrible pain in the stomach are soon followed by vomiting and diarrhea. Dehydration can come quickly, then convulsions and coma. Death ensues within hours. Chronic arsenic poisoning is much harder to detect. If the victim is given small doses in food over a period of time, arsenic can cause kidney and liver failure, which would not be associated with poisoning. Little wonder that arsenic was commonly referred to as "inheritance powder."

What did Nathan Straus, one of the founders of Macy's department store, refer to as "the white peril"?

Milk. Straus had become convinced that something had to be done about drinking unsanitary milk, which he thought was responsible for the high death rate among New York children. He, of course, was not alone in this belief. Many doctors and scientists had denounced the poor quality of milk and had linked it to disease. Straus, sensitized by the death of two of his own children, decided to take matters into his own hands. He was probably prompted by the death of his own cow, which an autopsy determined had died from tuberculosis. Had the cow passed the disease to his family? Straus wondered. Louis Pasteur's theories about disease and microbes were already well known at the time, and it was suspected that tuberculosis was a bacterial disease. And Pasteur of course had shown that bacteria could be destroyed by a brief heat treatment.

At his own expense, in 1893 Straus built his own pasteurization plant and set up a series of milk distribution centres in New York from which he sold his pasteurized milk for only a few cents. People who could not afford even that got it for free. Although Straus's venture is credited with saving thousands of lives, he was not unopposed. Activists claimed that pasteurization destroyed the nutritional value of milk and that the dead bacteria were more dangerous than live ones, and organized anti-pasteurization demonstrations. Straus did not capitulate and carried out a vigorous campaign to promote pasteurization, withdrawing not until 1920, when the United States government finally made pasteurization a requirement.

An interesting footnote to the story is that Straus devoted himself to philanthropic ventures totally as the result of the sinking of the *Titanic*. His beloved brother Isidore and Isidore's wife perished in the disaster, and a distraught Nathan retired from business to devote himself to philanthropy. His work was well appreciated. One admirer remarked in 1923 that Straus was "a star in the milky way of philanthropy, a man whose heart is bubbling over with the sterilized milk of human kindness."

⚘

What causes the often-seen green discolouration in potatoes?

Exposure to light. The green tint often seen near the surface of potatoes is due to chlorophyll, the pigment in plants that makes photosynthesis possible. Essentially, when potatoes are exposed to excessive light, they will try to photosynthesize; in other words they will try to sprout and grow into potato plants. The green colour is completely harmless, but another substance, called solanine, forms at the same time as chlorophyll. The theory is that this natural toxin forms as the potato tries to protect itself against pests and fungi during the critical sprouting process. Solanine, which falls into the family of compounds known as glycoalkaloids, can be toxic to humans. One would have to eat a fair amount of green potatoes to experience serious toxic effects, but even small doses can cause stomach problems. It is best to peel away the green, which will remove most of the solanine, which tends to concentrate near the surface of the tuber. The best way to avoid discoloration and potential health problems is to keep the potatoes out of the light and store them at a cool temperature. The production of solanine involves an enzymatic reaction that proceeds slowly at low temperatures. So don't worry about a little green discoloration in your potatoes, but do not go out of your way looking for green potatoes to eat.

※

What is the connection between the following drugs: Baycol, Propulsid, Seldane and Redux?

These medications were all approved after they had gone through the usual battery of safety and efficacy trials but were eventually removed from the market when problems cropped up. When a medication is removed from the market because of unacceptable side effects, the public often blames the company for having released it in the first place, and lawsuits inevitably follow. But the fact is that no amount of pre-market testing can guarantee safety, and side effects that occur very rarely will not be noted until the drug is in common usage. Baycol (cerivastatin), a cholesterol-lowering drug, was removed because at high doses it caused muscle damage. Propulsid (cisapride) was used for nighttime heartburn but was associated with a risk of fatal heart rhythm abnormalities. Seldane (terfenadine) was an antihistamine also linked to heart rhythm problems. And Redux (dexfenfluramine) was a weight-loss drug linked to heart valve abnormalities. In each of these cases the complication rate was extremely rare, but since alternative safer drugs were available, the manufacturers chose to remove them from the market. Such problems cannot be foreseen, and the eventual removal of a drug from the market does not mean that someone has been negligent. Of course, there have also been cases in which some drug companies have withheld negative information in order to try to keep a drug on the market. It's not a perfect world.

♀

A repairman working on a refrigerated soft-drink machine in a Texas hospital accidentally caused a small explosion that released a poisonous gas, causing a partial evacuation of the facility. What was the gas?

Phosgene. The refrigerant in the soft-drink machine was Freon, a substance that poses essentially no health risk but which has become notorious for playing a role in the destruction of the ozone layer when released into the environment. But when Freon is heated to a high temperature in the presence of a catalyst, such as copper, it decomposes to yield phosgene, a highly poisonous gas. This is exactly what happened when the repairman triggered a small fire in the vending machine. The copper pipes in the machine served to catalyze the formation of phosgene from Freon. Phosgene is an insidious poison because symptoms can take more than twenty-four hours to appear. Fluid fills the lungs, and the lung tissues begin to decompose. This happens because the gas reacts with moisture to yield carbon dioxide and hydrochloric acid, and the acid damages the lungs. The Germans were the first to put phosgene to a nefarious use by adding it to chlorine gas in order to increase the latter's killing "efficiency." Indeed, phosgene, not chlorine, was responsible for most of the 100,000 gas-caused deaths in the First World War. Luckily, in the case of the hospital accident, one of the firefighters called to the scene recognized the hay-like odour of phosgene and ordered the evacuation, which prevented any injuries. In other instances, injuries have occurred. Freon is a commonly used refrigerant on fishing vessels, and at least in one recorded case, it caused a serious injury. An accidental discharge of approximately 40 pounds (18 kilograms) of Freon in a commercial fishing vessel's engine room resulted in the formation of phosgene gas, causing severe lung burns in a crewman. Firemen have to be aware of the possible formation of phosgene any time they are called to a scene where Freon refrigerants may be present.

Workers who pick which fruit have to wear rubber gloves to prevent their hands from being eaten away?

Pineapple. Pineapples contain an enzyme called bromelain that breaks down proteins. The stem has a particularly high content of this enzyme, and when it is handled will actually result in a breakdown of the proteins in flesh. Indeed, if meat is marinated in fresh pineapple juice, it virtually falls apart. The activity of the enzyme also explains why you cannot add fresh pineapples to Jell-O. Bromelain breaks down gelatin, the protein responsible for the gelling effect. When bromelain is ingested, it is mostly broken down in the digestive tract, but some is absorbed into the bloodstream. Various claims have been made about the possible beneficial effects of ingesting purified bromelain, ranging from anti-tumour and anti-inflammatory effects to circulatory improvement and help with digestion. It is difficult to rationalize such effects theoretically, because even if bromelain makes it into the bloodstream, it is likely to be degraded very quickly by enzymes that protect the body from foreign proteins. Still, some controlled studies have shown benefits. In one case, boxers recovered from bruises more quickly when given bromelain, and in another, administration of 400 to 1,000 milligrams a day resulted in the improvement of angina symptoms. The overall evidence for health benefits is pretty sketchy, but bromelain does merit further investigation.

🔦

Perhaps the most dramatic case of poisoning in Canadian history occurred in 1964, when forty-eight men in Quebec became sick and twenty died after ingesting cobalt sulphate. How did this happen?

The men were beer drinkers, and cobalt sulphate had been added to the beer to make the foam last longer. This was a legal additive, but in 1964 breweries increased the amount they were using, hoping to increase sales. But all they did was increase misery. At first, the death of the men was a mystery. Only when investigation showed that the common link among the victims was the fact that they had each been drinking more than 200 ounces (6 litres) of beer a day for more than twenty years did authorities begin to suspect something in the beer. Researchers gave cobalt sulphate to rats and showed that it could result in death. Interestingly, though, cobalt caused death only if the animals' diet was lacking vitamins and protein. It seems that protein can prevent cobalt from being absorbed. This also explains why many people who drank the cobalt-containing beer didn't experience any effects. Only those who had a diet deficient in protein and vitamins and were heavy beer drinkers were poisoned by the cobalt sulphate.

It all goes to show that individually additives may be shown to be safe in test animals that are well nourished, but there may be risks if other nutritive factors come into play. Testing additives in healthy animals in sterile laboratories may not be realistic because humans have various ailments and live in a polluted environment. Furthermore, additives may react with each other, with prescription drugs or with pesticide residues. Testing all such combinations is impossible, so that declaring an additive to be safe is really just an educated guess. There are also idiosyncratic reactions. In rare cases sodium fluoride can cause loss of coordination, sleepiness and palpitations, as can tartrazine. MSG can cause severe headaches.

ॐ

What would happen to you if you bothered a whip scorpion?

You would be sprayed with a concentrated solution of acetic acid. This little creature isn't a true scorpion, although it looks like one.

It doesn't sting, but protects itself by spraying an 84 percent solution of acetic acid at attackers. This arachnid is also called the "vinegaroon," but that is also a bit of a misnomer because vinegar is only a 5 percent solution of acetic acid. The spray also contains some caprylic acid, a very foul smelling compound that promotes penetration of the acetic acid through tissue. Don't pick up a vinegaroon to look at it closely. The little guy has great aim and can let loose its chemical spray many times in succession.

$$\text{🍂}$$

Why might Ukrainian president Viktor Yushchenko be interested in including American "fat free" potato chips in his diet?

To reduce the concentration of a potent toxin in his body. Yushchenko was poisoned by the highly toxic chemical dioxin, which was somehow introduced into his food. He developed liver and pancreatic inflammation as well as a type of disfiguring facial condition known as chloracne. The inflammation subsided, but there is concern that he is now at risk for soft tissue sarcoma, a type of cancer linked to dioxin exposure. Dioxin is fat soluble, concentrates in fatty tissue and is difficult to eliminate from the body. This is where Olestra comes in. This non-absorbable replacement for fats was developed by Procter and Gamble. You can use it for frying, it has the mouth feel of fat and it even sort of tastes like fat. But you don't have to worry about weight gain, because it isn't absorbed—it is totally eliminated! Unfortunately, in some cases the elimination comes with a touch of diarrhea.

It turns out that Olestra, being a fat-like substance, dissolves dioxin readily. So any dioxin in the gut ends up being eliminated with the Olestra. But how do you get dioxins out of fat cells, into the blood and then to the gut? By losing weight. When fat is released from fat cells, the dioxins they harboured are also released and wend their way into the gut. If there is nothing there to absorb them, they get reabsorbed into the blood. But if Olestra is present in the gut, it traps the dioxins and eliminates them from the body. So as well as consuming Olestra, Viktor would have to cut down on his caloric intake. Researchers at the University of Cincinnati School of Medicine have shown that the concept works—at least in mice. In their experiment, weight loss along with ingestion of Olestra caused elimination of hexachlorobenzene, a fat-soluble toxin. There is even some human evidence about the benefits of using Olestra to eliminate fat-soluble toxins. A man who worked with electrical transformers developed toxicity to PCBs, a type of oil once commonly used in such equipment. Two servings of Olestra products a day for two years dramatically reduced the PCB level in his fatty tissues, and he made a remarkable recovery. Who would have ever thought that Pringles could be therapeutic?

❦

According to a proposal by California governor Arnold Schwarzenegger, the state will buy what item from public-spirited citizens?

Gasoline-powered lawnmowers. California is very much concerned about air pollution, to which gasoline-powered mowers contribute significantly. A single mower emits more pollutants than forty-three cars each year. It is estimated that retiring four thousand mowers would reduce air pollution by 20 tons per year, which is more than is generated by oil refineries in Los Angeles in just two days. One of the main pollutants spewed out by any gasoline engine is a family of compounds called aromatic hydrocarbons. These are known carcinogens. The amount of polycyclic aromatic hydrocarbons released in an hour of lawn mowing compares to that emitted by a car during a 90-mile (150-kilometre) trip. And whoever is pushing the mower is breathing in all these compounds. Fitting lawnmowers with catalytic converters would cut down on these emissions, as would switching to electric mowers. But the best idea is to use an old-fashioned push mower: no carcinogens and plenty of exercise!

☙

Gasoline station attendants have a surprisingly high rate of accidents driving home from work. Taiwan eliminated this problem with a law passed in 1997. What did the law require gas stations to do?

Gas stations had to install vapour recovery systems that prevent gasoline fumes from being vented into the air during fill-ups. We've all smelled gasoline as we fill up, and I think most people have a suspicion that the inhaled vapours are not doing them any good. They are right. As usual, though, toxicity is a matter of dose, and inhaling the vapour once a week during a fill-up is probably of little consequence. But gas station attendants are exposed to the fumes all day. That seems to be a different matter. Taiwanese researchers found that attendants who pumped gas had on their way home from work an accident rate two and a half times higher than office workers. There was no difference in accident rates during the drive *to* work. Proof that gasoline vapours were responsible for this effect came after passage of the 1997 law. The difference in accident rate between the gas station attendants and the office workers disappeared. Vapour recovery systems suck gasoline fumes back into the storage tank during fill-up and can cut vapour release by 90 percent. Installing these is an effective way to reduce air pollution, health problems—and car accidents.

Death sentences in the United States are carried out by either electrocution or cyanide gas except in cases where the condemned criminal has agreed to donate his organs. What form of execution is used in such instances?

Injection of a high dose of potassium chloride. Potassium is a great example of a substance that is beneficial in small doses but lethal in larger ones. A small amount of potassium, in the form of potassium ions, is essential for life. It is required for proper functioning of nerve cells, the kidneys as well as muscle tissue, including the heart. But an overdose causes the heart to stop beating. Unlike electrocution or cyanide, potassium does not damage any of the organs that are candidates for transplant. Potassium deficiency is uncommon because numerous foods contain the element in its absorbable ionic form. Raisins, nuts, peanuts and bananas are particularly good sources. Patients who are prescribed diuretics to lower their blood pressure by increasing urine output can lose significant amounts of potassium and are often counselled to increase their intake of these foods.

The name of potassium derives from potash, the ash left behind when wood burns. In 1807 Humphry Davy passed an electric current through a moistened batch of potash and observed that a shiny metal deposited around the negative electrode. He had isolated metallic potassium in a pure form from potash.

❡

What country has the highest per capita death rate from snakebite?

Sri Lanka. About a thousand people are killed every year by snakes in Sri Lanka, which has a population of roughly twenty million. Not only are there lots of snakes, but Sri Lanka also harbours the world's most dangerous snake, the Russell's viper. Named after a Scottish physician who a couple of hundred years ago wrote a guide to poisonous snakes, this viper injects its victims with a deadly mix of chemicals. Some of these interfere with blood coagulation, which in turn leads to brain hemorrhage and kidney failure, while others cause muscles to break down. One in every ten people bitten dies. The snake preys on rodents that commonly live in rice paddies and tries to avoid people. But around harvest time the paddies fill with people, and the snakes are disturbed and strike back. An antivenom is available, but it is not very effective because it is based on the venom of the Indian Russell's viper, which is not exactly the same as the Sri Lankan one. About forty thousand people succumb to snake bites in the world every year.

❡

A 2006 directive from the European Union aimed at reducing pollution caused great concern among builders of church organs. What were they worried about?

The regulation was designed to limit potentially hazardous substances such as lead, mercury and cadmium to 0.1 percent by weight in any product that works on electricity. The reason for the new law is that discarded cell phones, computers and other electronic equipment may release toxic metals into the environment. Church organs come under the law because modern organs use electric fans to blow air through the pipes, which are made mostly of lead. Organs that use the old-fashioned bellows pumped by foot action are not affected. While the EU is concerned about the manufacture of new organs, it is also funding a project to prevent historic organ pipes from crumbling.

Ctesibus of Alexandria is usually credited with inventing the organ in the third century BC, but it took about a thousand years before the instrument became a mainstay in churches. When Columbus returned triumphantly from America in 1493, a Mass celebrated in his honour featured organ music. These instruments produce their delightful tones when air is blown through pipes of different lengths. The organist controls the amount of air streaming through the pipes with multiple keyboards and foot pedals. Although historically some pipes were made of wood, most were made of lead with small amounts of other metals mixed in. Pipes in which tin was alloyed with lead suffered from a condition known as tin plague. Tin crumbles into a powder at low temperatures, which then causes the disintegration of the pipes. One suggestion attributed tin plague to the devil's trying to silence God's music. Eliminating tin is not the answer, because this metal actually prevents lead from undergoing another type of corrosion.

Lead develops holes when it reacts with acids in the organ wind. This obviously interferes with the quality of the music.

And where do the acids come from? Wood, particularly oak, which was widely used in organ construction, emits acetic and formic acids as it ages. The problem was exacerbated as churches introduced central heating, and over many years the acids eventually ended up pitting the lead. The higher the concentration of tin in the alloy, the less severe this particular problem. Seems you just can't win.

The research project funded by the EU aims to find ways to slow corrosion and repair damaged organ pipes. Sometimes amputation is the only resort, and sections of pipe have to be cut out and replaced. The effort is certainly a worthwhile one because organs represent both a scientific and cultural heritage that needs to be preserved. What a tragedy it would be if some of the magnificent organs that show the artistry of ancient masters were allowed to corrode away. And as for manufacturers of the new electrically powered organs that use lead pipes, the EU says that companies can apply for an exemption from the law. After all, their products are not likely to end up in landfills.

What was the first insecticide to be extracted from a plant?

Nicotine. Tobacco had been poisoning people for centuries before the juice of the plant was used as an insecticide in the seventeenth century. Tobacco plants produce nicotine to protect themselves from predators, and applying the compound to other plants affords protection for these as well. But such application must be carefully done because nicotine is highly toxic and can be absorbed through the skin. About 60 milligrams is enough to kill an adult if it gets into the bloodstream. Back in 1940, a woman in Britain mixed nicotine into her husband's aftershave and killed him. That amount of nicotine can be extracted from a couple of cigarettes. Although cigarettes contain a potentially lethal amount of nicotine, most of it burns before it is inhaled. Even eating a few cigarettes is not a problem because absorption of nicotine from the digestive tract is very slow. Chewing tobacco is actually more dangerous because absorption through the thin membranes in the mouth is easier than through the stomach wall or the intestines.

Perhaps this is where Georgette Heyer got the idea for the deadly toothpaste in *Behold Here's Poison*. In that novel, the murderer extracts nicotine from tobacco and uses a syringe to inject it into the victim's toothpaste. Ed McBain in *Poison* was even more inventive. In this story a jealous dentist obtains nicotine from a lab doing tests on stained teeth and hides some of it in a temporary crown covering a root canal, making sure the crown has a thin spot that will be worn away by chewing or brushing. Charlotte MacLeod's *Bilbao Looking Glass* features a victim killed by gulping a martini dosed with nicotine. Far fetched? Perhaps not. In 1968 a woman did in her elderly sister by mixing the residue of cigarette butts in a jug of water, straining it and placing the water at her bedside. Today, nicotine is still used to kill aphids on roses.

In terms of the number of people affected, Chernobyl was the largest nuclear radiation accident in history. What was the second largest?

In 1987 hundreds of people in Goiânia, Brazil, were contaminated when scavengers broke open a cylinder they had found in an abandoned cancer clinic. The cylinder turned out to contain radioactive cesium chloride.

Early radiation therapy devices used cesium-137 as a source of gamma rays to destroy rapidly dividing cancer cells. In 1971, such a machine was imported from the United States by a private clinic in Goiânia but was supplanted by a more modern cobalt-60 device in 1978. When the clinic moved in 1985, the cesium machine was left behind in the abandoned building. A couple of years later two scavengers going through the building were intrigued by the machine and removed the lead cylinder that contained the cesium source. After much effort they managed to break it open and were stunned by the beautiful blue light that emanated. This effect results from the absorption of gamma radiation by chloride ions, which then re-emit some of the energy as visible light.

Thinking they had something valuable on their hands, the scavengers took the cylinder to a local scrap dealer, who bought it for the equivalent of about twenty-five dollars. Workers at the scrapyard then dismantled the cylinder and were amazed to find a glowing substance inside. Thinking it had some sort of magical properties, they shared it with co-workers, who took some home to show off the marvellous material. Children played with it, and one man even used the glowing dust to paint a cross on his abdomen. It wasn't long before people began to show up at clinics with skin burns, falling-out hair and other signs of radiation sickness. Within a short time 244 persons were found to be contaminated and 54 had to be hospitalized. Four of these soon died, including a six-year-old who had rubbed the powder on her body so that she glowed and sparkled. One of the original scavengers had to have his arm amputated.

The long-term outcome of the disaster is still not known because the effects of radiation may take a long time to reveal themselves. Cesium can replace potassium in the body and become incorporated into tissues and bone, where it can trigger cancer. Leukemia rates have already risen in the area.

As can be imagined, there was a great deal of panic among the population, as exemplified by the protests against burying the six-year-old victim in a local cemetery. This worry was not totally unjustified, as her body was indeed radioactive. She now lies in a Goiânia cemetery in a lead coffin surrounded by concrete, an unfortunate reminder of the largest nuclear radiation accident in the Western Hemisphere.

ℙ

A Canadian member of Parliament made the following comment: "It is not okay to put poison in our food just because it's properly labelled." What poison was he referring to?

Trans fats. Pat Martin certainly meant well and undoubtedly hoped to improve the health of Canadians by suggesting that foods that contain more than 2 percent trans fats should not be marketed. But referring to trans fats as "poison" is going overboard. Singling out any food component as a devil, or indeed as a saviour, is scientifically unsound. Food is an incredibly complex mix of chemicals, and at best we can talk about healthy diets and unhealthy diets. Trans fats are no more of a poison than the unavoidable saturated fats that are found in meats and dairy products. Although most trans fats are the result of the hydrogenation process used to convert liquid oils into solid fats for the food processing industry, trans fats also occur in nature. Beef fat, for example, contains about 2 percent trans fats, and the fat from lambs has about 5 percent. The concern about trans fats is that, like saturated fats, they increase the risk of heart disease. There is evidence for this, but it is not quite as ironclad as people think.

Several European studies have failed to link heart disease with trans fats, and the Nurses' Health Study in the United States, which followed more than ninety thousand nurses since 1976, found only a slight connection between heart disease and fat consumption. Interestingly, in North America, over the last couple of decades, while trans fat consumption has stayed constant, the rate of heart disease has declined significantly. That being said, it is still a good idea to reduce trans fat intake, particularly because trans fats are a marker for foods with poor nutritional value. We can certainly do with fewer cookies, cakes, danishes, crackers, doughnuts and french fries. But let's make something clear. Eliminating trans fats from such foods does not make them "healthy." Our aim should be to eat a truly healthy diet. After all, when you are eating whole grains, fruits, vegetables and lean meat, you don't have to worry about trans fats.

There is another problem with legislation against trans fats. Determining the amounts present in food is not so simple. That's why foods that have less than half a gram per serving can legally state that they contain none. The fact is that trans fat concentration cannot be determined accurately to half a gram. In any case, following Canada's Food Guide—advice that is less emotionally charged than calling for the elimination of trans fats—will automatically limit the intake of these substances that Mr. Martin calls "toxic poison." I am certainly not opposed to eliminating trans fats from our food supply. There is even precedence for workable legislation. In Denmark, for example, foods containing more than 2 percent trans fats cannot be sold, and the Danish food industry has not collapsed. But let's not jump to the conclusion that eating a danish for breakfast is fine in Denmark but is toxic in North America. Just eat oat bran topped with ground flax and berries instead. This is healthful on both sides of the ocean.

☙

Carbon monoxide is the product of incomplete combustion, and when inhaled is responsible for numerous cases of poisoning every year. But how is it possible to be poisoned by carbon monoxide without being exposed to the gas?

Carbon dioxide can form in the body when certain solvents, such as paint strippers, are metabolized in the liver. The liver is the body's main detoxicating organ, breaking down intruders into molecules that can be more readily eliminated from the body. But the process isn't perfect, and sometimes the breakdown products turn out to be toxic themselves. Such is the case with methylene chloride, a solvent found in many paint strippers and in industrial degreasing solvents. When paint stripper is inhaled in a poorly ventilated environment, it can cause nausea, dizziness and headaches, but an even greater danger is metabolism in the liver to carbon monoxide. The antidote is oxygen inhalation, but if the situation is untreated it can be fatal.

The effects of carbon monoxide can occur after exposure to the solvent has ended. This may seem strange, but it is explicable. Methylene chloride is fat soluble, and after it is absorbed into the bloodstream from the lungs, it distributes around the body and dissolves in fatty tissue. Then later it emerges from the fat and is metabolized in the liver to carbon monoxide. This is not a theoretical scenario. Carbon monoxide poisoning can be recognized because it turns mucous tissue a cherry red colour, and this has certainly been seen after exposure to methylene chloride. Just another reason to be very careful with solvent inhalation.

$$\mathcal{Q}$$

Irène Joliot-Curie, Marie Curie's daughter, like her mother, died of leukemia. To what was her leukemia attributed?

In 1941 a capsule containing polonium-210 exploded on her laboratory bench, exposing her to alpha radiation, which is thought to have caused her death fifteen years later. Polonium was first isolated from pitchblende, a uranium ore, by Marie and Pierre Curie in 1898. Uranium was known to be radioactive at the time, but the radioactivity of pitchblende was greater than what could be accounted for by uranium. The Curies eventually isolated a tiny amount of a new radioactive element, which they named polonium, after Poland, Marie Curie's homeland.

Isolating the polonium was no small task. It required starting with several tons of pitchblende to isolate less than a milligram of polonium. Today, polonium-210 can be made in gram quantities by bombarding bismuth with neutrons in a nuclear reactor. It is extremely toxic because of the alpha radiation it emits. It is a trillion times more toxic than hydrogen cyanide, meaning that it is lethal in quantities as small as a picogram—a millionth of a millionth of a gram. Since alpha particles do not penetrate the skin, polonium has to be inhaled or ingested to be lethal.

Polonium can be regarded as a natural carcinogen. Radon is a gas that occurs in nature and radioactively decays. One of the decay products is polonium, which can then lodge in the lungs as tiny solid particles. Small amounts of polonium-210 are used on brushes to remove static electrical charge from photographic film. The alpha radiation ionizes the air through which it passes, and the resulting ions neutralize the static electrical charge.

$$\mathfrak{P}$$

In 1974 the U.S. Food and Drug Administration asked that hairsprays containing which ingredient be removed from the market?

Vinyl chloride, which was used as a propellant. Spray cans require the use of a substance that under ordinary conditions is a gas but can readily be liquefied when pressurized. When the nozzle on the can is opened to the atmosphere, the propellant quickly evaporates, and as it escapes from the container it propels the active ingredients along with it. Obviously another requirement is that the propellant not react with any of the other ingredients. Vinyl chloride met the required criteria, and in the 1950s and '60s was commonly used in various products, including insecticides and hairsprays. The substance was readily available because it was widely used to produce polyvinyl chloride, or PVC—one of the most useful plastics ever developed. PVC was used to make all sorts of consumer items, ranging from toys and shower curtains to water pipes and coverings for car seats. As early as 1938 there were reports of toxicity to vinyl chloride, the monomer that was used to make PVC. Workers in the PVC-manufacturing industry reported dizziness and confusion when exposed to vinyl chloride vapours. And then in 1949 liver damage was reported in fifteen of forty-eight workers exposed to vinyl chloride in Russia.

By 1971 the type of liver damage in workers had been clearly identified, and the news was not good. Vinyl chloride caused angiosarcoma of the liver, a rare but extremely dangerous form of cancer. The industry began to implement measures to curb workers' exposure to vinyl chloride, but there is still controversy over the time frame involved. Some industry documents show that companies were aware of the vinyl chloride risk before it became a matter of public record and were slow to respond to the danger because limiting workers' exposure could interfere with production. In 1974, however, the U.S. government declared angiosarcoma of the liver to be an occupational disease after six cases were diagnosed among vinyl chloride workers at the B. F. Goodrich Chemical Company plant in Louisville, Kentucky. This forced the industry to adopt stricter controls over the amount of vinyl chloride that workers could inhale. At that time the FDA also asked that spray cans containing vinyl chloride be withdrawn, and the era of using vinyl chloride as a propellant came to a halt. It is virtually impossible to know the risk the public experienced, but it would be interesting to know the health status of hairdressers who in the 1950s and '60s used hairsprays liberally. They were likely exposed to levels of vinyl chloride known to cause cancer in animals.

Today, the standard allows vinyl chloride in workroom air at 1 part per million during an eight-hour period, and the industry has come up with equipment that complies with this. Consumers do not have to fret about using PVC products, as these do not contain any of the vinyl chloride monomer.

§

What chemical released from permanent press fabrics has been linked with skin rashes?

Formaldehyde. Not many people like ironing. But they don't like wearing wrinkled clothing either. And they do like the feel of cotton. Unfortunately, cotton and wrinkles go together, at least until a little chemical ingenuity comes into the picture. Cotton is composed of long molecules of cellulose that are linked to each other by weak bonds referred to as hydrogen bonds. The rungs of a ladder would be an appropriate analogy. When cotton fabric becomes wet, water disrupts these bonds, and the cellulose molecules can move relative to each other. This random movement is what causes wrinkles. As the fabric dries and water evaporates, adjacent cellulose molecules re-form the hydrogen bonds, but these now hold the fabric in the new, wrinkled shape. Treating the cellulose molecules in the original fabric with chemicals that link them together with bonds that cannot be disrupted by water results in a fabric that will retain its shape. Essentially, the rungs of the ladder have been made more durable.

Various chemicals are available to form these "cross-links" between cellulose molecules, the most popular of which have been urea-formaldehyde resins, used since the 1950s to impart wrinkle resistance to garments. They work well, but raise a concern about the release of small amounts of formaldehyde, which can cause contact dermatitis, better known as a skin rash. While this is a relatively rare reaction as far as the general public is concerned, it is an issue for workers in the textile industry. Handling fabrics that have been treated with formaldehyde resins has been linked with skin problems, and more significantly, formaldehyde exposure has also been associated with headaches, burning eyes, and nose and throat irritation. There is also the lingering concern that formaldehyde causes cancer in laboratory animals. As a consequence, chemists have been hard at work trying to find cellulose cross-linking agents that minimize formaldehyde exposure. They have managed to come up with dimethyloldihydroxyethylene urea, or DMDHEU, which may twist your tongue, but not your health. Formaldehyde release from this compound is virtually negligible. Formaldehyde-free reagents have also been developed, but they are not free of problems. They either are very expensive or cause discoloration of the fabric. Unfortunately, all cross-linking treatments weaken the fabric, an effect referred to in the trade as "crosslink embitterment." And so the search goes on to find improved ways of making fabrics wrinkle resistant, long wearing and formaldehyde free. Of course, if you are concerned about the chemicals used in such processes, you can always avoid permanent press and go back to the ironing board.

Why was Agent Orange so called?

The mixture of herbicides known as Agent Orange was shipped to Vietnam in drums marked with orange stripes for recognition. Some 21 million gallons (80 million litres) were sprayed over three and a half million acres (1.4 million hectares) in Vietnam and Laos between 1962 and 1971, mostly to reduce enemy hiding places by defoliating trees. Agent Orange was a mixture of two chemicals, 2,4,5-trichlorophenoxyacetic acid (2,4,5-T) and 2,4-dichlorophenoxyacetic acid (2,4-D), both of which had been developed in the

1940s to rid agricultural fields of weeds. The compounds resemble the molecular structure of a plant hormone known as auxin and can mimic its effects. When sprayed on broad-leaf plants, 2,4-D and 2,4,5-T induce very rapid growth, resulting in the death of the plant. Neither 2,4-D nor 2,4,5-T are very toxic to humans, but a contaminant that was present in 2,4,5-T turned out to be terribly dangerous. Known as tetrachlorodibenzodioxin, or just dioxin for short, it was capable of causing birth defects and cancer in animals at incredibly low doses.

There is no question that millions of Americans and Vietnamese were exposed to Agent Orange and consequently to dioxin. How they might have been effected is something that has been debated for decades. The U.S. government has maintained that disease patterns in Vietnam veterans are no different from the rest of the population, while veterans' organizations claim to have data that proves otherwise. Overall, the epidemiological evidence suggests that some cancers are more prevalent in populations exposed to dioxin. In 1979, a large class-action lawsuit was filed against the manufacturers of Agent Orange, and was settled out of court in 1984. The result was the Agent Orange Settlement Fund, which distributed nearly $200 million to veterans between 1988 and 1996. Claims, counterclaims and legal hassles continue to this day.

It should be understood that dioxin was not an intentional component of Agent Orange. One of the chemicals used to manufacture 2,4,5-T, namely 2,4,5-trichlorophenol, can undergo a side reaction and produce the notorious tetrachlorodibenzodioxin. On account of this possibility, 2,4,5-T was phased out by the late 1970s. The other component of Agent Orange, 2,4-D, is still widely used around the world as a weed killer. Since 2,4,5-trichlorophenol is not involved in its manufacture, there is no way it can be contaminated with tetrachlorodibenzodioxin. There have been some claims that 2,4-D itself may be linked to cancer in farmers who are extensively exposed, but there is no reputable evidence to suggest that the use of 2,4-D in products such as Killex for home lawn care presents a risk.

☙

The murder victim died from a heart attack when a burst of adrenalin released glucose from glycogen stored in his liver. With what had he been injected?

Insulin. Insulin is a hormone produced by the pancreas and is required by cells other than those of the nervous system for the proper utilization of glucose, the body's prime energy source. Type I diabetics cannot produce insulin and require regular injections of the substance. But the amount of insulin injected has to be balanced against the food consumed and the extent of physical activity. Food of course provides glucose, and activity uses it up. This balancing act is not easy, and overdoses of insulin do occur. When this happens, glucose in the bloodstream is forced into muscle cells, leaving an insufficient amount in the blood to support the energy needs of the brain.

Hypoglycemia, or low blood sugar, is a condition that requires immediate attention. Diabetics know that if they overdose on insulin they must immediately consume some form of sugar to avert a serious reaction disaster. Since for the body the sugar needs of the brain are a priority, as blood levels of sugar drop, a message goes out to release glucose from glycogen, the form in which it is stored in the liver for emergencies. This message is transmitted by adrenalin, released from the adrenal glands. But such a burst of adrenalin can also trigger a heart attack, especially in the old and the young. This is the reaction that makes insulin a possible murder weapon.

There have been a number of celebrated criminal cases in which insulin has been implicated. Perhaps the most famous was that of Claus von Bülow, who in 1982 was accused of killing his socialite wife, Sunny, with an insulin injection. He was convicted, but later tried again when evidence inconsistent with the insulin murder came to light, and von Bülow was freed. Beverly Allitt, on the other hand, is spending her life in prison. This British nurse, whom the press called the Angel of Death, killed four children and attempted to kill five others in her care in the 1990s at a hospital in Lincolnshire by injecting them with insulin. Allitt was thought to suffer from Munchausen by proxy syndrome, a condition in which people deliberately injure others in their care to bring attention to themselves. She had craved attention all her life, and even going to prison did not stop her attempts to attract attention. Allitt repeatedly stabbed herself with paper clips and poured boiling water on her hand. Obviously her brain doesn't function properly, but in this case, not because of any lack of glucose.

What contaminant in Hungarian paprika caused a national scandal in 1994?

Lead. Hungarian kitchens ground to a halt in 1994, when sales of paprika were suspended. What happened? Red oxide of lead had been detected in a third of paprika samples tested by government inspectors! This was more than just a nasty bit of consumer fraud; forty people were hospitalized with lead poisoning. Eighteen people were eventually arrested before Hungarian eyes and palates could again be satisfied. Paprika is central to Hungarian cuisine, both for its flavour and its colour. It is made by grinding up dried capsicum peppers, originally introduced into Europe from America by the Spaniards. There are many varieties of these peppers, with the extent of the zing they provide being determined by the amount of capsaicin they contain. That's the same compound that effectively stops rioters when used in the controversial pepper sprays. Depending on capsaicin content, paprika can be either sweet or hot.

While the flavour comes mostly from capsaicin, the colour of paprika is due to various pigments known as carotenoids, of which lycopene is the dominant one. That's the same lycopene found in tomatoes, the compound we've heard so much about in connection with the prevention of prostate cancer. The reddest paprika contains the most lycopene and is the most appealing visually. That is what led to attempts to improve the colour of paprika artificially, as happened in the Hungarian scandal. Under the Communist regime, paprika production, like everything else, was under government control. There were two state-owned paprika mills and they produced paprika according to strictly prescribed government standards. But with the introduction of free enterprise in the late 1980s, paprika mills mushroomed. Competition became fierce, and the notion of improving the colour by clandestine means became appealing to some producers. The scandal had the effect of stimulating strict regulations on paprika production, ensuring that Hungarian paprika today is the best in the world.

mysterious
connections

What contribution did William Congreve and Lt. Henry Shrapnel make to baseball?

Baseball games just wouldn't feel right without the singing of "The Star-Spangled Banner," whose "rockets' red glare, the bombs bursting in air" are familiar to just about everyone. And therein lies the connection. William Congreve was responsible for the red glare, and Shrapnel for the bombs bursting in air. The American national anthem was composed by Francis Scott Key, one of the defenders of Fort McHenry near Baltimore when it was attacked by British frigates during the War of 1812. The attack featured a bombardment of the fort by rockets launched from the ships. By today's standards these rockets were primitive, but they were certainly fearsome in 1812. Their range was more than a mile and they exploded in the air, raining shrapnel on those below. In those pre-antibiotic and pre-antiseptic days, shrapnel wounds would often cause infections leading to death. The risk of such wounds eventually resulted in the introduction of the now commonplace metal helmets worn by soldiers.

The term *shrapnel* derives from the name of the British lieutenant who devised a shell that would explode above the ground and spread its fury. Originally these shrapnel shells were fired from cannons, but then William Congreve introduced his rocket. His idea was not original; Congreve had witnessed British forces being attacked by troops in India equipped with rockets. He improved upon these by attaching a stabilizing stick so they could be better targeted and fitted them with shrapnel bombs. "Congreve rockets" were made in various sizes, but the 32-pound variety were most commonly used. Both the rocket and the shrapnel shell were powered by black powder, which was a mix of charcoal, sulphur and saltpetre. When the rockets streaked through the sky, their exhaust left a red glare, and then their shrapnel bombs exploded in the air. This is what Francis Scott Key commemorated in "The Star-Spangled Banner."

Incidentally, the "flag [that] was still there" still is there, in Washington at the National Museum of American History. The flag is 50 feet (15 metres) long but is somewhat tattered. Workers are painstakingly restoring it by hand to fix the damage caused by the efforts of William Congreve and Henry Shrapnel.

What is the relationship between the Lone Ranger and Argentina?

Everyone knows, or should know, that the Lone Ranger used silver bullets. But what may not be common knowledge is that the country of Argentina derives its name from argentum, the Latin name for "silver." It is in fact the only country in the world named after an element. South America is rich in silver deposits, and at the time of the Spanish conquest in the sixteenth century, the natives were already knowledgeable about the extraction of the metal from its ores. The Spaniards plundered the natives' silver, shipping it down a river before transporting it to Spain. The river came to be called Rio de la Plata, or River of Silver. In English today it is River Plate. When people of the area were finally able to free themselves of Spanish rule in the nineteenth century, they named their new country Argentina after its mineral wealth.

Now, back to the Lone Ranger. Why did he use a silver bullet? It is lighter than lead, and with the same amount of gunpowder it travels faster. It is also harder than lead and will shatter a bone more readily. Obviously, it's also great for knocking a gun out of the bad guy's hand. And how much faster does the silver bullet travel? At a distance of 15 feet (4.5 metres), just about the distance for a quick-draw gunfight, about 2 milliseconds. Not much of an advantage. But the Lone Ranger didn't have to rely on the speed of his bullet. He was a good guy, and good guys always win. He'd knock the gun out of the criminal's hand (good guys don't hurt people), jump on his horse and be on his way with a cry of "Hi-yo, Silver!" Then the following week he would return to the same big rock and do it all over again.

Why do collectors of antique golf balls have a special appreciation for root canals?

Gutta-percha, a rubber-like material from the sap of sapodilla trees of East Asia, was once used to make golf balls and is used today to fill the root canals of teeth after diseased pulp has been removed. It is a harder material than rubber and more resistant to water. The difference in properties between gutta-percha and rubber comes down to a question of molecular structure. Both substances are composed of giant molecules, in other words, polymers, built of repeating isoprene segments. But the isoprene monomers are joined together in a slightly different way, and that causes the difference in properties. (For the chemically astute: the double bonds have the *cis* configuration in rubber and *trans* in gutta-percha.)

Now, for the golf balls. Yes, there are people who collect them, and they love to get their hands on any made roughly between 1850 and 1900. That's when gutta-percha balls replaced leather balls stuffed with feathers. The earliest "gutties" were handmade from pieces of gutta-percha that were softened by heating in water. Eventually a skilled gutty maker was able to produce six dozen or more balls a day, twenty-five times what a feather-ball maker could turn out. This meant that the price of golf balls dropped, and consequently lower-income golfers could afford to play the game. But today, low income earners can't afford to collect gutta-percha golf balls. Although some can be found for a few hundred dollars, hand-hammered balls produced between 1855 and 1875 have been auctioned off for as much as $28,000. For that, you can even get a root canal. And when you get one, you'll note the smell. That's isoprene being released when gutta-percha is heated. It may even smell familiar. Human sweat also contains isoprene, from the breakdown of vitamin A. I suppose I could have asked what the relationship was between root canals and sweat, but that could have another answer.

What is the connection between the expression "saved by the bell" and Edgar Allan Poe?

Poe was captivated by accounts of people who supposedly had been buried alive. His story "The Premature Burial" deals with this frightening possibility, albeit somewhat tongue-in-cheek. In the 1800s there really was concern about the possibility of being buried alive, based on stories going around about coffins being dug up with scratch marks on the inside of the lid. In response to this, some coffin designers arranged for a rope connected to a bell above the ground. If the supposed dead came back to life, the bell would be rung, and if someone were around to hear it, the unfortunate victim could be saved. "Saved by the bell," that is! Well, they had urban legends back in the 1800s too. There is no evidence that any coffins were dug up with scratch marks, although coffins really were removed and the bones buried elsewhere to make room for the newly dead. In any cases, scratch marks would have been invisible on the rotted wood.

The idea that the expression "saved by the bell" comes from the fear of being buried alive is nonsense. It originates from the sport of boxing, in which a fighter can be saved from a knockout by the bell that signals the end of a round.

♀

What is the common link between gunpowder, vulcanized rubber and acid rain?

The element sulphur. Gunpowder is a mixture of sulphur, charcoal and potassium or sodium nitrate, better known as saltpetre. When sulphur burns, it reacts with oxygen to produce a gas, sulphur dioxide. In gunpowder the oxygen is supplied by the saltpetre, and it is the sudden expansion of the sulphur dioxide gas as well as carbon dioxide produced by the burning charcoal that propels the projectile. The ancient Chinese and Arabs produced primitive incendiary mixtures, but it was Friar Roger Bacon in England around 1245 who found that the best mix was 75 percent saltpetre, 15 percent charcoal and 10 percent sulphur. Roughly at the same time, Berthold Swartz in Germany came to the same conclusion. In any case, gunpowder was first used in Europe at the Battle of Crécy in 1346, at which Edward III defeated the French. Sulphur was known in biblical times as brimstone, and the book of Genesis describes how "the Lord rained upon Sodom and Gomorrah brimstone and fire."

But it was long after this, in 1839, that American inventor Charles Goodyear found another use for sulphur. He was looking for a way to improve the properties of natural rubber, which tended to become soft and sticky at high temperatures and stiff and brittle when it was cold. In fact, Goodyear became so obsessed with solving this problem that he ended up in jail on several occasions for not being able to repay people the money he had borrowed for his research. Then one day a lucky accident occurred. Goodyear had been trying to mix rubber with sulphur to see what would happen and spilled the mix on a hot stove. Presto, vulcanized rubber was born, named after Vulcan, the Roman god of fire. It turned out that the sulphur atoms formed links between the long chains of rubber molecules, making it less sensitive to temperature. Today, we drive around with tires made of vulcanized rubber. And we burn gasoline that contains small amounts of naturally occurring sulphur. When this burns it reacts with oxygen and moisture in the air to form sulphuric acid, which is a major component of acid rain. Of course the problem extends beyond automobiles. Any time we burn oil or coal, which always contain some sulphur, we produce acid rain. That is undesired, but the industrial production of sulphuric acid from sulphur is highly desirable. In fact sulphuric acid may be the world's most important industrial chemical. It is essential for the production of fertilizers and is commonly used to make a myriad of products, ranging from paints and explosives to steel and detergents.

$$\text{\Large ♀}$$

What is the link between Neil Armstrong and coloured mouthpieces for trumpets?

The helmet that Neil Armstrong wore when landing on the moon in 1969 and coloured trumpet mouthpieces are made of the same plastic, Lexan. In 1953 Daniel Fox at General Electric was looking for better insulating materials for wires when he mixed together a couple of ingredients that he hoped would yield a polymer with suitable properties. What he got was a goop that hardened to such an extent that he couldn't even remove his stirring rod. Banging the curious new material against a hard surface had no effect; it just couldn't be broken. Lexan, as the new material came to be known, would never be used as wire insulation, but it would find a host of other uses. It was as clear as glass but virtually unbreakable. Electric meter covers were traditionally made of glass and often made inviting targets for rambunctious rock-throwing teenagers. But Lexan meter covers solved this problem. Lexan was also ideal for astronauts' helmets, automobile headlight assemblies and panels for greenhouses. Uses mushroomed.

Football and hockey helmets, protective visors, laptop housings, iPods, cell phones, water bottles, bulletproof shields, skateboards, airplane canopies, compact discs and DVDs joined the list of items made from Lexan. And then along came plastic mouthpieces for trumpets and trombones to serve as an alternative to brass pieces. The plastic doesn't change temperature as easily as brass, providing more comfort for musicians' lips. It is also cheaper and comes in a variety of attractive colours.

Chemically, Lexan is a polycarbonate, made by linking together molecules of phosgene and bisphenol A in an alternating fashion. Since polymerization never goes to 100 percent completion, trace amounts of bisphenol A are always left over in Lexan. Some of this can escape from the plastic, to the concern of environmentalists. They worry that bisphenol A has estrogenic properties and can have a disruptive effect on human hormones. This concern is not totally unreasonable, since some experiments have shown that animals exposed to tiny amounts of bisphenol A experience physiological effects. No such effects have ever been found in humans. In any case, as with any such issue, it comes down to a risk-benefit evaluation. Life as we know it, with our computers, iPods and plethora of sports equipment, would be drastically altered if Lexan were removed from the market. Police would be deprived of their shields, athletes of their helmets, fighter pilots of their protective canopies. Real lives would be lost for the sake of protecting ourselves from the theoretical risk attributed to trace amounts of bisphenol A leaching out from Lexan products.

What is the connection between the 2005 film *Charlie and the Chocolate Factory* and artificial tears?

The movie used fake chocolate made with the same thickening agent, hydroxyethyl cellulose, that is used to make artificial tears. The movie's script called for a chocolate river and waterfall, but using such massive amounts of real melted chocolate was of course out of the question. That's why the producers approached Vickers Laboratories to solve the problem of making over 260,000 gallons (a million litres) of fake chocolate that would look like the real thing, would be safe for actors to frolic around in and would last throughout the filming. The chemists decided on using water thickened with food-grade hydroxyethyl cellulose, colouring it with organic pigments and adding a food-grade biocide as a preservative.

Thickening involves adding substances to water that impede the movement of water molecules. Long molecules, or polymers, do an excellent job. Starch is a classic substance used to thicken sauces; it disperses readily and gets in the way of the free movement of water molecules. It has the added benefit of forming so-called hydrogen bonds with water, further restricting the movement of the H_2O molecules. Hydroxyethyl cellulose acts the same way. Less than 1 percent in water has a remarkable thickening effect. That's why it is used in shampoos, conditioners and various creams. In artificial tears, the thickened liquid sticks readily to the eye and lubricates it.

Vickers actually built a plant near the movie's set to produce the huge amount of fake chocolate needed to supply the river and waterfall. It looked chocolatey enough and was safe for the actors to swim in. But they didn't get much pleasure from this—even if they swallowed some of the stuff, there was no taste.

Chemists were not the only scientists needed for the movie. Animal trainer Michael Alexander and his team spent nineteen weeks training forty squirrels for the squirrel-room scene. The squirrels were trained to sit on stools while they opened nut shells and dropped the nuts onto a conveyor belt.

♀

What is common to Thomas Edison, Leonardo da Vinci, Albert Einstein, Woodrow Wilson and Winston Churchill?

They all had dyslexia. Judging by these names, it is immediately obvious that dyslexia is certainly not a form of mental retardation. Dyslexia is a type of learning disability characterized by difficulties with direction, as in left-right or east-west, or with spelling and reading. It is usually diagnosed when a disparity is noted between a child's intelligence and actual achievement. Dyslexics are often very intelligent and speak well but have difficulties with reading and spelling. Churchill did so poorly in school that his father once exclaimed: "I have an idiot for a son." But Winston was no idiot. With special coaching he was able to enrol in the Royal Military College, and the rest, as they say, is history.

♀

What is the connection between the word *salary* and oven cleaner?

Our word *salary* derives from the Latin word *sal*, which means "salt." The ancient Romans recognized the importance of salt for preserving food as well as for tanning leather, and actually paid their soldiers with it. Today, salt is an extremely important raw material in the chemical industry for the production of chlorine and sodium hydroxide. These chemicals are produced by passing an electric current through a salt solution. Sodium hydroxide is the active ingredient in most oven cleaners. It readily reacts with fatty substances, converting them into water-soluble "soaps." Indeed, commercial soap is made by reacting fats with sodium hydroxide, also known as lye.

What is Christmasy about Iscador, a reputed anti-cancer drug?

The supposed active ingredient in Iscador is an extract of the mistletoe, which of course is used as a decoration during Christmas festivities. The idea of using a mistletoe extract to treat disease can be traced back to Rudolf Steiner, a mystic who was interested in various alternative therapies. He was taken by the fact that mistletoe is a parasitic plant, meaning that it grows not in the soil but in the bark of trees. It grows perpendicular to the branch in which it thrusts its sucker, and it does not obey many of the laws of the plant kingdom. For instance, its berries ripen in winter, without warmth. It stores up chlorophyll and is green all year long and is indifferent to light. And does Iscador cure cancer? No. But surprisingly, there is some evidence that Iscador when used in conjunction with conventional treatments does enable patients to live longer. More studies are needed before specific recommendations can be made.

$\wp$

What is involved in both the greenhouse effect and remote-control devices for home entertainment?

Infrared radiation. Infrared radiation is that part of the electromagnetic spectrum that we sense as heat and is made up of a range of wavelengths much like visible light. In the case of visible light, we recognize the different wavelengths by their different colours. As with visible light, infrared light can be separated into its component wavelengths by passing it through a prism. Remote-control devices generate specific wavelengths to which sensors in TVs and other such equipment respond. Now for the greenhouse effect. Visible light from the sun passes through the earth's atmosphere and warms the earth's surface. The warm surface emits infrared radiation, but unlike visible light, this does not pass through the gases in the atmosphere. Some of the infrared radiation is absorbed by molecules such as carbon dioxide, methane and water vapour in the atmosphere. When these molecules absorb infrared radiation they become energized and twist and turn and move about more quickly. This is what we sense as an increase in temperature. Water vapour is actually a far more effective greenhouse gas than carbon dioxide, but its concentration in the atmosphere is pretty constant, whereas the concentration of carbon dioxide is increasing as a result of human activity.

Although we use the term greenhouse effect to describe the warming of the atmosphere, it turns out that this is not a very good analogy.

The term was introduced in 1896 by the famous chemist Svante Arrhenius, who believed that glass in a greenhouse acted like carbon dioxide in the atmosphere. However, as early as 1909, an experiment with two greenhouses, one covered with glass and the other covered with rock salt, which does not absorb infrared radiation, showed that the temperatures reached inside were almost the same. In a greenhouse the increase in temperature mostly stems from a lack of ventilation.

Something else we should remember about the warming of the atmosphere. The greenhouse effect is only part of the problem. Heat is a pollutant we cannot avoid. Any time we produce energy, some of it winds up as heat. Power plants that burn coal or oil not only produce carbon dioxide but also dump vast quantities of residual heat into the environment. Nuclear power plants, which do not produce carbon dioxide, still produce heat as a side product. When we run our cars, ovens, washing machines, dryers, we produce heat. Every manufacturing process, be it that of plastic bags or paper produces some heat. René Dubos, the renowned biologist, put it well, although perhaps in a bit of an extreme fashion: "We will destroy our lives by producing more useless, destructive energy to make more and more needless things that do not increase the happiness of people."

What is the link between marshmallows and cattle bones?

Gelatin. Marshmallows are made of sugar and gelatin, and gelatin is a protein that is derived from cattle bones. Gelatin is typically produced in a powdered or granulated form. Slightly yellow to light tan in colour, it is a rather tasteless and odourless substance.

Gelatin use in the food industry is probably best recognized in gelatin desserts and confectionery applications such as marshmallows and gummi candies. It is also used as a binding and glazing agent in meats and aspics. Because it is highly digestible, in the pharmaceutical health industry gelatin is used to make the shells of hard and soft capsules for medicines, dietary and health supplements, and syrups. The unique chemical and physical properties of gelatin make it an important component in the photographic industry as well. Gelatin serves many useful purposes in the preparation of silver halide emulsions in the production of photographic film. A new, major application for gelatin is in the paintball industry. The classic-style "war games" are played out using projectiles constructed of gelatin.

The raw materials used in the production of gelatin are from healthy animals and include cattle bones, cattle hides and fresh, frozen pigskins. In the North American market, these raw materials are sourced from government-inspected meat processing facilities.

What chemical connection is there between the bombardier beetle and the classic *I Dream of Jeannie* TV show?

Hydrogen peroxide. The bombardier beetle has a fascinating defence mechanism. When in danger, it sprays its enemy with a hot mix of chemicals. The beetle stores hydrogen peroxide in a compartment in its belly and when needed mixes it with a catalyst to produce oxygen and water. This reaction releases heat and converts the water into steam that forces the mixture out with explosive force.

In *I Dream of Jeannie*, the same chemistry was used when Barbara Eden emerged from her bottle in a puff of white smoke. The white smoke was really steam, generated by mixing hydrogen peroxide with the catalyst manganese dioxide in the bottle.

♔

What chemical connection is there between ancient Chinese ink and modern car tires?

Carbon black, the soot obtained by burning organic compounds. The ancient Chinese discovered that burning tree resins or vegetable oils produced a fine black powder that could be used to make watercolours and inks. The Greeks and Romans commonly made black paint from carbon black, and it is still widely used today to make pigments. The toner in our computer printers owes its colour to carbon black. It is also the stuff that is responsible for the colour of our automobile tires. Here, though, it is not colour we are after, it is strength. Back in 1904, S. C. Mote, the chief chemist of the India Rubber, Gutta Percha and Telegraph Works Company, discovered that carbon black had reinforcement properties in rubber, and by 1912 it had replaced zinc oxide as the primary reinforcing agent in tires. The process that made tires possible in the first place was vulcanization, invented by Charles Goodyear. Heating natural rubber with sulphur hardened and toughened the material, but it was the eventual addition of carbon black that dramatically improved performance.

Today, tire manufacturing is a sophisticated process, with various types of rubber and additives being used, but carbon black is still the essential ingredient. Numerous varieties have been developed. Regular tires, for example, use a coarse grade of carbon black, while racing tires use a much finer grade. The fine grains have a large surface area, and extensive rubbing between the grains produces a high temperature through internal friction. As the tire warms up, the rubber becomes more sticky and provides more traction. In fact, maximum grip is obtained at around 190°F (90°C), which is why we often see race-car drivers swerve their vehicles violently during the warm-up lap. The friction created by the sudden directional changes warms up the tires. Stickiness of the rubber is great for grip, but not so good for mileage. Of course race-car drivers don't worry about this, but everyday drivers do. That's why regular tires use a different type of carbon black. Racing-tire manufacture is a very secretive business because in many cases races are won or lost depending on tire performance. And this performance is a reflection of the type and particle size of carbon black that is used.

☢

Why did California issue a health advisory about traditional Mexican treats such as candies and fried grasshoppers?

Worry about contamination with lead. Chili powder is at the heart of Mexican cuisine and finds its way into numerous foods, ranging from candies and salsa to fried grasshoppers. Unfortunately, sometimes lead finds its way into the chili powder, occasionally in amounts high enough to cause toxicity, especially in children. This was discovered when routine testing of foods in California revealed high levels of lead in particular candies sold in Hispanic neighbourhoods. These spicy candies, imported from Mexico, were made with guajillo chilies, but the source of the lead was a mystery. An investigation was in order because in some cases the lead content was way more than the permissible level of 0.2 parts per million. Lead is an insidious poison—even small amounts are capable of causing kidney damage, behavioural problems and learning disabilities. Children are especially sensitive because lead accumulates in their still-developing skeletal and nervous systems.

The problem of lead toxicity eased with the banning of lead-based household paints in 1978 and leaded gasoline in 1986, but other environmental sources of lead remain, such as the unusual case of chili peppers. Where was the lead coming from? Investigators found two sources. When chilies are dried, they are spread out on concrete slabs, where they are exposed to dust and dirt. Many soils contain naturally occurring lead compounds, which may become a problem if they are concentrated. That is just what happened in the case of the peppers, which were never cleaned before being ground up into chili powder. But a secondary source was even more disturbing. Since farmers sell their chili peppers in burlap sacks to millers by weight, they sometimes add scrap metal, including lead, to increase the weight. The millers will use magnets to remove the scrap, but lead is not attracted to magnets. Chilies destined for foreign sales are cleaned, but apparently producers take many shortcuts with the local Mexican market, and it was candies produced for local consumption that found their way across the border into California. Hence the warning about candies and fried grasshoppers. Salsa was never a danger because in the manufacturing process the chili is sufficiently diluted.

Every time you turn on your home oil furnace or drive around in your car, you make the writing on some gravestones less legible. What accounts for this effect?

Acid rain. Many gravestones are made of marble, which dissolves in acid. This means that if rainwater is acidic, every time there is some precipitation, a bit of the marble is worn away. Acid rain can mostly, but not exclusively, be traced to human activity. The fossil fuels that we rely on extensively for heating and transport are the end product of the long-term decomposition of plants and animals in the soil. The carbohydrates, fats and proteins that once made up these living creatures are converted into the hydrocarbons that make up the bulk of petroleum. But many protein molecules also contain sulphur atoms, and these remain as a contaminant in fossil fuels. When the fuels burn, the sulphur reacts with oxygen to form sulphur dioxide, which in turn combines with moisture in the air to form sulphuric acid, and hey presto, we have acid rain! Cars make a secondary contribution to acid rain as well. The high temperatures inside the engine allow nitrogen and oxygen, the major components of air, to react and form oxides of nitrogen, which then react with water to produce nitric acid. Nature, too, makes a contribution here, as lightning also produces the high temperatures that allow nitrogen and oxygen to react.

Now back to our gravestones. Both marble and limestone are made of calcium carbonate, a base that can neutralize acids. As this happens, though, the calcium carbonate is converted to calcium sulphate or calcium nitrate, which is then washed away. Basically, every time it rains, a bit of marble gravestones, or indeed limestone buildings like Canada's Parliament, are worn down. This kind of destruction, though, is not the only problem with acid rain. It increases the solubility of toxic metals such as copper and lead, so that leaching from water pipes becomes more extensive. It can also damage the leaves of trees, as the maple syrup industry has discovered. Sap production has reduced as industrialization and traffic have increased. Fish populations in acidified lakes decline, which in turn means that animals that depend on aquatic ecosystems are also affected. Birds that feed on fish start to go hungry.

Not all lakes are equally affected by acid rain; the damage depends on the kind of bedrock present. Limestone neutralizes acids and has a buffering effect, but granite, composed of silicates, does not undergo any acid–base reaction. So a lake surrounded by granite rock is likely to suffer more damage from acid rain. But granite's lack of reaction with acid also means that gravestones made of granite are going to last longer than those made of limestone.

🎠

What are potassium acetate, sodium chloride, calcium chloride, magnesium chloride, and urea all used for?

Melting snow and ice on roads. All of these substances interfere with the formation of ice crystals and can be used to melt ice. They do, however, differ in effectiveness, potential harm to the environment and cost. Sodium chloride, or common salt, is cheap and can melt ice down to about -5°F (-20°C). But it can also damage soil and vegetation, contaminate surface and groundwater and speed up the corrosion of concrete and metals. Corrosion of metals is a process whereby the metals react with oxygen. This requires the transfer of electrons among substances, and such transfer is facilitated by the presence of ions such as sodium and chloride. Substances that dissolve to form ions in solution are called electrolytes and speed up the rusting process. That's why cars in Canada rust and those in Arizona do not. That is also why airplanes not in use are stored in the Arizona desert, where there is virtually no humidity.

Not all electrolytes speed up corrosion to the same extent. Potassium acetate is much more environment friendly than salt but is twenty times more expensive. Calcium chloride melts snow and ice much faster than sodium chloride and is less corrosive, but does damage vegetation and wildlife. It also costs more than salt. Magnesium chloride is also less corrosive, but works only down to 5°F (-15°C) and costs five times more than salt. Urea is non-corrosive and does not damage vegetation, but melts ice only down to 25°F (-4°C). There is yet another issue with salt. It works more effectively if the grains can be prevented from clumping. To reduce caking, small amounts of sodium ferrocyanide are added. In the presence of sunlight, this compound can break down and release cyanide, which can be washed into waterways and damage aquatic life. Ethylene glycol or propylene glycol are the substances used to de-ice airplanes because they are non-corrosive. These liquids are collected and recycled, but some is inevitably lost to the environment. Obviously there is no perfect way to melt ice and snow. But not using these substances would result in loss of life. As with so many scientific issues, it is a question of evaluating risk versus benefit.

What is the connection between the red colour of highway warning flares and a town in Scotland?

The element strontium is responsible for the redness of flares and is named after Strontian, a Scottish town where the mineral from which it was isolated was discovered. In 1878, Adair Crawford, a Scottish-Irish chemist, became interested in a display of the mineral witherite in a Scottish museum and asked for a sample to study. Witherite is mostly barium carbonate, but on examination Crawford realized that the sample also contained a second mineral that had not been noted before. He named the new mineral strontianite. In 1807, Humphry Davy reacted the mineral with hydrochloric acid and produced strontium chloride. One of Davy's research interests was the behaviour of substances when subjected to an electric current. Electrolysis of strontium chloride yielded a shiny soft metal that could not be decomposed further and was therefore identified as an element. Strontium metal has little commercial use, but the fact that its compounds can produce a red colour in flames has importance in flares and fireworks.

Some accounts suggest that Buddhist priests in ancient India amazed their flock by generating brilliant red flames in temples. The theory is that they achieved the spectacular effect by mixing coal, sulphur and an oxidizing agent such as potassium chlorate with strontium compounds. This chemistry is very similar to what happens in modern fireworks. But strontium has a less benign connection as well. It is one of the elements that forms when uranium undergoes fission, as in a nuclear reactor or in an atom bomb. The issue here is that the specific isotope of strontium that forms, namely strontium-90, is radioactive, and is present in the fallout after a nuclear explosion or in an accidental discharge from a nuclear power plant. Strontium can then settle on grass or hay that is eaten by cows, and the element gets concentrated in milk. This presents a health hazard to humans drinking the milk because chemically strontium is similar to calcium and gets incorporated into bone, where the energetic beta particles that the radioactive strontium emits can damage bone marrow and cause cancer. Of course, this is a concern only when a fission reaction occurs. No radioactive strontium is present in flares or fireworks. So it is safe to live in Strontian, Scotland. At least as far as radioactivity is concerned.

♀

What product contains cream of tartar, sodium bicarbonate, sodium aluminum sulphate and starch?

Double-acting baking powder. The purpose of baking powder is to generate carbon dioxide gas and cause batter to rise as it is being baked. Sodium bicarbonate, or baking soda, reacts with acids to release carbon dioxide, as anyone who has ever mixed the stuff with vinegar knows. The key to baking is to allow the carbon dioxide to evolve continuously throughout the baking process. When baking powder is first dissolved in the batter, the cream of tartar, which acts as a mild acid, immediately reacts with the bicarbonate to release carbon dioxide. This reaction keeps going until all the cream of tartar has reacted. Sodium aluminum sulphate also acts as an acid when dissolved in water but does not become active until a higher temperature is reached. The result is an evolution of gas throughout the baking process. Starch is included in the dry product to absorb any moisture and prevent premature loss of carbon dioxide.

What property, other than potential toxicity, do mercury and bromine share?

These are the only two elements that are liquids at room temperature. Bromine is a brownish liquid with a boiling point of 138°F (59°C), and evaporates readily to form a brown vapour. The liquid is very corrosive and will inflict a nasty burn if spilled on skin. The vapour is highly irritating when inhaled. Mercury is a shiny metallic liquid with a boiling point of 675°F (357°C), which means that it does not vaporize readily. The liquid is not corrosive and can be readily handled. In fact, if it is ingested, it just goes right through the system. Mercury vapour, on the other hand, is toxic. Even though the vapour pressure of the liquid is not high, over time it does evaporate enough to create a problem. A broken thermometer, if not cleaned up, can release enough mercury to affect people in a house, especially children. That's why mercury thermometers are being phased out. Compounds of mercury, such as methylmercury, are more toxic than the element itself. This is the form of mercury that is present as a contaminant in some fish, although to a very small extent. Four other elements, francium, cesium, gallium and rubidium, become liquids at temperatures slightly above room temperature.

What substance derives its name from a large gypsum deposit in the Montmartre area of Paris?

Plaster of Paris, or calcium sulphate hemihydrate. Gypsum, or calcium sulphate dihydrate, was originally obtained from rock quarried at Montmartre, a suburb of Paris. Heating gypsum drives off water and yields plaster of Paris. When the dry plaster powder is mixed with water, it re-forms into gypsum, initially as a paste but eventually hardening into a solid. Plaster is used as a building material similar to mortar or cement. Like those materials, plaster starts as a dry powder that is mixed with water to form a slurry which can readily be applied to surfaces. Unlike those materials, plaster remains quite soft after drying, and can be easily manipulated with metal tools or even sandpaper. These characteristics make plaster suitable as a finishing, rather than a load-bearing, material.

This common construction material indirectly led to the first law governing food adulteration in England, in 1860. Adding plaster of Paris to flour and sugar was commonplace in those days in order to extend these expensive commodities with a cheap adulterant. Unfortunately, one day a druggist's boy was instructed to add plaster of Paris to a batch of peppermint lozenges but reached into the wrong bin and added white arsenic, a rat poison, instead. Thirty people died, but the tragedy precipitated the passage of the Food and Drug Adulteration Act. In a bizarre twist, calcium sulphate was one of the first food additives approved under the new act. We still use it today in bread making to provide yeast with calcium, an essential nutrient for these microbes. And that is why rye bread today is plastered!

What flower was named for its resemblance to testicles?

The orchid. About two thousand years ago, Dioscorides, a Greek medical writer, noted that the bulbs of the orchid looked like testes. The Greek word for "testes" is orchis, from which of course we derive our word orchid. The ancient Greeks believed that elixirs made from orchid roots would have aphrodisiac qualities and be a cure for various reproductive problems. Of course orchids do not possess any such property, but they are indeed fascinating flowers. Certain orchids ensure that their pollen gets spread around by engaging in sexual activities with bees. Female bees attract males by producing a variety of straight-chain hydrocarbons. Well, some orchids mimic this scent and also produce flowers that are shaped like the female bee. The male bees then mistake the orchids for female bees and engage in pseudocopulation with the flowers, causing pollination. The Catasetum orchid has different-looking male and female flowers. When a bee lands on a male flower, it gets a nasty surprise: it gets sprayed with a sticky mass of pollen. It then avoids males and seeks only females, carrying the male's pollen from flower to flower.

What link is there between diabetes and eating fish?

Mercury, a contaminant is some fish, may destroy the cells in the pancreas that make insulin. The connection is by no means ironclad, but a study at the National Taiwan University raises a red flag. First of all, how does mercury get into fish in the first place? Mercury compounds occur naturally in coal, and when coal is burned in power plants the mercury is released into the atmosphere and then gets washed down by rain into water systems. Some industries that manufacture chlorine from salt also use large amounts of mercury, and mercury-based light switches were used in cars until 2003. Some of this mercury escapes when cars are scrapped. Mercury is also used as a catalyst in the production of PVC plastics, in dental amalgams and in some gold-mining processes. Once mercury enters water systems, bacteria convert it to its most hazardous form, namely methylmercury. This is the substance that builds up in fish.

Methylmercury has strong oxidant properties, and it is because of this that it destroys cells. The Taiwanese researchers showed that this was the case by adding an antioxidant, N-acetyl cysteine, to their cell cultures and observing that it protected the cells from damage by methylmercury. It is important to remember that the effects of methylmercury were seen in the lab, not in living systems, but the amount of methylmercury that caused damage was typical of what is found in contaminated fish. A follow-up study has, however, shown that mice fed low levels of methylmercury for a month do produce less insulin and have higher blood glucose levels. There is no comparable evidence in humans, and in fact there is a question of whether or not methylmercury is absorbed into the bloodstream. While mercury, and possible other pollutants, may contribute to diabetes, there is no doubt that the dramatic increase being seen in North America is due to the increase in obesity.

♀

What consumer product is named after a famous Scottish surgeon?

Listerine. In the late 1800s, Joseph Lister sought an explanation for why fractures that broke through the skin usually would become infected whereas those that did not pierce the skin healed nicely. The prevailing opinion at the time was that the exposed tissues were affected by oxygen in the air. Oxygen would break down the molecules of organic material in a wound, turning them to pus. In an attempt to exclude oxygen, the common practice was to dress the wound with tight bandages. Doctors even resorted to collodion, a solution of gun cotton in ether and alcohol that formed a film as the solvents evaporated. Actually, these tight bandages encouraged bacterial growth and resulted in a virtually indescribable stench in the wards. In fact many doctors believed that the stench caused the infections and was directly responsible for the roughly 50 percent death rate after surgery. Yet, incongruously, nobody tried to solve the problem by eliminating the smell. The sole light in the darkness belonged to that pioneer of modern nursing and "the lady with the lamp," Florence Nightingale, who espoused a regimen of soap, warm water and sunshine but was largely ignored.

💡

What type of commercial product resulted from a scientist making an interesting observation about the area around a bottlebrush?

A weed killer. The bottlebrush is an ornamental bush that grows in temperate climates and has flowers that resemble a brush used to clean bottles. In 1977 a scientist working for the Stauffer agrochemical company (eventually to become Syngenta) noted that no weeds grew around the bottlebrush plants in his back yard. This observation stimulated research into the chemistry of the plant, and researchers soon discovered that the bottlebrush released a natural herbicide into the soil. Leptospermone, as the compound came to be called, was an allelopathic substance, meaning that it was produced by a plant in order to harm another plant. In this instance, the effect was to kill nearby weeds. The commercial potential for such a substance is obvious: weeds are the enemies of crops, and their control has huge economic significance. Extraction of leptospermone from bottlebrush was difficult, but the compound was amenable to synthesis. When tested in the laboratory, it turned out to be effective against a variety of weeds, but the required application rate was too high to make leptospermone commercially viable.

In such cases, the usual procedure is to make some alterations in the molecular structure of a substance to increase activity. Literally thousands of variations were tested until, eleven years after the initial isolation of leptospermone, Syngenta chemists came up with mesotrione, which turned out to be a very effective weed killer, particularly in corn fields. Mesotrione, like leptospermone, interferes with an enzyme plants use to synthesize carotenoids, which protect chlorophyll from destruction by excess light. The result is that chlorophyll is destroyed, photosynthesis stops and the plant dies. Typically weeds become "bleached" as they lose their green chlorophyll. Corn is unaffected by mesotrione as it quickly breaks the compound down into inactive metabolites. Mesotrione, sold commercially as Callisto, is not toxic to humans and biodegrades quickly. An interesting example of a casual observation by a scientist leading to an important discovery.

What invention links Michael Faraday, Thomas Edison and James Watt?

The electrical generator. Michael Faraday was intrigued by electro-magnets, which had been invented by British electrician William Sturgeon, who in 1825 had taken a horseshoe-shaped piece of iron, around which he wound a coil of wire. When he passed an electric current through the wire, the metal became magnetized. Joseph Henry, an American, made a far more powerful electromag-net a few years later and demonstrated the potential of Sturgeon's device for long-distance communication by sending an electronic current over one mile of wire to activate an electromagnet that caused a bell to strike. This was the idea behind the telegraph. Realizing that an electric current could create magnetism, Faraday wondered if magnetism could create an electric current. In 1831 he discovered that moving a magnet inside a coil of copper wire caused a small electric current to flow through the wire. This was the world's first electric generator.

Edison grasped this idea and brought it to fruition. His main strength was to take laboratory curiosities and turn them into practical devices. Edison succeeded in making an electric generator that could produce enough power to illuminate his lab and then managed to scale it up to light a few streets in New York. But to generate enough elec-tricity to light cities, more powerful generators were needed. This is where Scottish engineer James Watt's steam engine came in. Watt had not invented the steam engine, but he had improved simple, inefficient models to a point that made the generation of electricity a practical proposition . . . practical proposition. Although Watt had nothing to do with lightbulbs, it was this invention that gave him ever-lasting fame. Virtually every lightbulb is rated in terms of watts, the unit of electric power named after the man whose steam engine made the gen-eration of electricity possible.

☙

What is the link between absinthe and malaria?

The wormwood plant. Actually the term *wormwood* refers to a number of perennial plants that are botanically related and share the
name *Artemisia*. One of these, *Artemisia absinthium*, is a traditional
component of absinthe. Its bitter taste contributes to absinthe's
distinctive flavour, and the plant also provides the chlorophyll that
gives the beverage its characteristic green colour. The psychoactive
properties of absinthe are thought to be due to thujone, which
occurs naturally in *Artemisia absinthium*. A related species, *Artemisia
annua*, produces artemisinin, a compound that has proven to be an
effective treatment for malaria, the terrible disease that has killed
about half the humans who have ever lived. The name malaria,
meaning "bad air," was coined by the ancient Romans, who thought
the disease was caused by vapours emanating from swamps. They
weren't completely wrong. Swamps do play a role, but not by releasing any sort of noxious vapours. A moist environment is the ideal
breeding ground for mosquitoes that transmit the malaria-causing
parasite when they feast on human prey. Today, over half a billion
people every year are stricken by malaria, and more than a million
of these, mostly children, die.

The most effective way to curb the ravage caused by the disease
is to control mosquito populations. DDT, the insecticide that
resulted in a well-merited Nobel Prize for its discoverer, Paul
Müller, was extremely successful in the 1950s and '60s at reducing
the incidence of malaria. Then along came Rachel Carson and her
classic book *Silent Spring*, which pointed out the risks of DDT to
falcons, eagles, sea lions and salmon. The risks were real, but were
not due to application of DDT against mosquitoes. Farmers trying

to protect their crops against insects overused DDT and tainted the soil and waterways with the compound. Concern eventually led to the banning of DDT in many parts of the world, a ban that, according to some experts, resulted in the deaths of millions of children. Today, the World Health Organization looks on DDT in a different light and encourages the application of small doses to the walls of buildings in areas where malaria is rampant.

The first effective treatment for malaria was quinine, extracted from the bark of the South American cinchona tree. When mosquitoes began to develop resistance to it, chemists synthesized a related compound, chloroquine. But chloroquine is also losing its effectiveness, and the search is on for better drugs. One of these is artemisinin, isolated from a species of wormwood. Amazingly, the anti-malarial action of wormwood was first described by the Chinese in the fourth century. Modern science did not pay much attention to this because the Chinese had also ascribed such properties to some two hundred other plants. But when desperate Chinese scientists screened these plants for their effectiveness, they found that only one, wormwood, had potential. And they were right. Today artemisinin is widely used as a component in anti-malarial medications.

chemicals in action

A man playing tennis on a hard court suddenly
noted bursts of flame where his shoes contacted the
surface. What was the cause?

Weed killer. Specifically, sodium chlorate—which also happens to
be a powerful oxidizing agent. This means that it has the ability
to release oxygen when it is heated. Basically, sodium chlorate
($NaClO_3$) decomposes to yield sodium chloride ($NaCl$) and oxygen. If there is any combustible material around when this reaction
occurs, it can catch fire. In the case of our tennis player, he was
wearing rubber-soled shoes. As he ran across the court, the friction
between the shoe and the surface generated enough heat to decompose the sodium chlorate and release oxygen. The rubber in the
shoe also served as fuel, which then ignited.

Such a reaction is of course undesirable on a tennis court but is
the essence of fireworks. These always include an oxidizing agent,
such as sodium chlorite, sodium chlorate or sodium perchlorate,
which release the oxygen required to ignite other components that
result in the colourful displays. Magnesium powder, for example,

produces brilliance as it burns to yield magnesium oxide, and aluminum flakes ignite to form luminous tails of aluminum oxide. Various colours are produced by including specific metallic salts. Strontium compounds result in reds, barium compounds yield greens, and calcium compounds produce orange colours. Blues are the toughest to achieve, usually with copper acetoarsenite.

Sparklers, which produce a shower of sparks, are based on the same type of chemistry. They contain an oxidizer, usually potassium chlorate or potassium nitrate, a fuel that burns, and a metallic powder that is ignited as the fuel burns to produce the sparks. The components are held together with a binder that also acts as a fuel. A typical sparkler, then, consists of potassium chlorate, powdered carbon (fuel), dextrin (a type of sugar as binder) and powdered aluminum and magnesium. These chemicals are blended into a slurry, and the sparkler is made by dipping a wire into the mix, slowly withdrawing it and allowing the mix to solidify. Obviously there are dangers any time that oxidizing agents are employed. Three sparklers can generate as much heat as a blowtorch. Every year, hundreds of people are admitted to hospital because of accidents involving fireworks.

§

Why does superglue set more quickly in Miami than in Phoenix?

It is all a question of humidity. A tube of superglue actually doesn't contain glue. It contains glue precursors, which, when exposed to moisture, turn into glue via the process of polymerization. Polymers are long chain-like molecules that can form when individual

units, called monomers, join together by means of a chemical
reaction. In the case of superglue, the monomers belong to a class
of compounds called cyanoacrylates, of which methyl-alpha-
cyanoacrylate is the most common. When cyanoacrylates are
exposed to moisture, they link up to form polycyanoacrylates,
which is the superglue. All surfaces are covered with a thin layer of
moisture, so as soon as the glue is applied it begins to polymerize
and harden into a solid mass. Miami is more humid than Phoenix,
so it isn't surprising that the glue hardens faster.

Cyanoacrylate adhesives can do more than stick surfaces together.
They can even catch criminals! That's because they can be used to
reveal latent fingerprints. If an object suspected of having fingerprints is placed into a chamber that contains cyanoacrylate vapour,
polymerization takes place where fingerprints are located, since the
prints always contain some moisture. The glue that forms then
sticks to the amino acids and fats that make up a fingerprint, resulting in a visible image.

Why do drug enforcement agencies become alarmed if they find that someone has made a large order of oil of sassafras?

Sassafras trees are commonly found in North America. The bark
and leaves have been sold as ingredients for "health" teas, but there is
no evidence they do any good. In fact, sassafras contains safrole, a
known carcinogen. Because of this, safrole cannot be used legally as
a food additive; at one time there was an intent to use it in root beer
flavouring. Safrole can also serve as the raw material for the synthesis

of illegal drugs, such as methamphetamine and ecstasy. If some-
body orders safrole from a chemical company, authorities take
note. For example, the Chicago Police Department launched an
investigation when someone claiming to be from Northwestern
University repeatedly ordered safrole from a major chemical pro-
ducer, So far, the culprit has not been apprehended.

☽

What cosmetic product has the same active ingredient as smokeless gunpowder?

Nail polish. The ingredient responsible for the formation of the
film that coats nails is nitrocellulose, the same substance that is
used to make smokeless gunpowder and various explosives.
Nitrocellulose is made by reacting cellulose from wood or cotton
with a mixture of nitric and sulphuric acids. When nitrocellulose is
dissolved in a solvent and the solvent is allowed to evaporate, a clear
film remains. Since nails grow every day, the film has to be flexible.
To achieve this, various substances known as plasticizers are added.
Camphor has plasticizing properties, but the most common plasticizers used have been dibutyl phthalate, dioctyl adipate and acetyl
triethyl citrate. Phthalates are, however, being phased out because
some have been linked with developmental problems in animals.

Various pigments can of course be incorporated, as well as compounds such as guanine, which is derived from fish scales, to yield
an appropriate lustre. Commercially this is known as natural pearl
essence. A cheaper way to produce lustre is to add bismuth oxychlo-
ride or mica, a mineral product. The pigments are finely dispersed
powders that do not dissolve in the nail polish; they are very well

dispersed so they do not affect gloss. They can be inorganic, such as iron oxides or organic dyes in the form of "lakes," which are usually aluminum or barium salts.

To improve the lacquer's sticking properties, resins made from tosylamide and formaldehyde are commonly added. Some concern has been raised about the possibility of formaldehyde causing skin sensitization, but the likelihood of this happening is low. While formaldehyde is present, it is not in the "free" form; it is chemically bound in the resin. Solvents for dissolving and dispersing the ingredients in nail polish are chosen from the likes of butyl acetate, ethyl acetate, acetone, amyl acetate, ethanol or toluene. A typical nail polish base consists of: nitrocellulose (13%), tosylamide/formaldehyde resin (11%), dibutyl phthalate (5%), ethyl acetate (22%), butyl acetate (41%), isopropanol (6%) and stearalkonium hectorite wetting agent (2%).

§

Superabsorbent disposable diapers are now available to be worn by young children in swimming pools. Instructions, however, clearly state that they cannot be worn in the ocean. Why?

The active ingredient in such diapers is a powder composed of polyacrylate, a polymer that has an amazing ability to absorb water. The result is a gel that firmly stays put in the diaper. These polyacrylates were originally developed for the space program in order to absorb any loose moisture droplets that might be floating around under weightless conditions. Unfortunately the gel is not stable in a salt solution and quickly breaks down, releasing the liquid it had held. The more salt in solution, the less effective the polyacrylates are in forming a gel. That's why the powder can absorb eighty times its weight of distilled water but only thirty times its weight in urine. Urine contains salt, but not as much as sea water. Sea water causes the gel to break down immediately, which is of course undesirable both for the child and anyone nearby.

$$℘$$

How do we know that the Tin Man in *The Wizard of Oz* was not really made of tin?

He rusted! In Frank L. Baum's book *The Wonderful Wizard of Oz*, when Dorothy asked the Tin Woodman what she could do for him, he replied, "Get an oil-can and oil my joints. They are rusted so badly that I cannot move them at all; if I am well oiled I shall soon be all right again." Just like a tin can, the Tin Man was misnamed. He was actually made of iron coated with a thin protective layer of tin. Iron is much cheaper than tin but it reacts with oxygen and moisture to form insoluble ferric oxide—rust. Tin has much less of a tendency to react with oxygen, but if there is a crack in the tin layer, the iron underneath will rust. This undoubtedly is what happened to the Tin Man.

By the time Baum wrote his classic, people were familiar with the tin can. It had been invented by Peter Durand, an Englishman, in 1810 as a response to a French discovery. Nicolas Appert had toiled for fifteen years to win a prize of 12,000 francs that the French government had offered to anyone who could come up with an effective way to preserve food. Appert had an idea. Wine did not spoil when tightly bottled, so perhaps neither would food. Eventually he found that partially cooking food, sealing it in bottles with a cork and then immersing the bottles in boiling water resulted in food that kept very well. But the Brits were not to be outdone by the French, and Durand came up with an iron can coated with tin that could be sealed. Unlike the French version, it did not break easily. By 1813, tins of food were being sent to British soldiers and sailors for trial. The tin cans passed with flying colours, and a year later were already being dispensed to British military bases, including the island of St. Helena, to which Napoleon Bonaparte was destined to be exiled. Ironically, it was Napoleon who had presented the 12,000-franc prize to Nicolas Appert for his invention of preserving food in glass jars. And now not only did the Emperor have to eat British food, he had to do it from British tin cans! Mon dieu!

🍾

MacGyver was a television hero who often resorted to ingenious science in his adventures. He once filled a crack in a chemical tank with chocolate, and the ensuing chemical reaction stopped the leak. What chemical was in the tank?

Sulphuric acid. Chocolate contains a lot of sugar, and sugar reacts with sulphuric acid to produce a hard foam that consists essentially of carbon. In the TV show, this foam successfully plugged up the leak in the tank of sulphuric acid. Would it really work? Well, that is sort of doubtful. The chemistry, though, is interesting. Sugar is a carbohydrate, and carbohydrates are so called because originally they appeared to be composed of carbon and water. Their molecular structure is actually much more complicated than that, but chemical reactions that remove water from carbohydrates do leave behind a residue of carbon. If sugar and sulphuric acid are mixed, an impressive but dangerous reaction occurs. As the white sugar changes to black carbon, a tremendous amount of heat is generated, and steam and droplets of sulphuric acid quickly evolve. Some of the steam is trapped inside the carbon, forming bubbles and creating a foam. A column of black foam impressively rises out of the mix. Nobody, except experts like MacGyver, should attempt this reaction because the hot sulphuric acid fumes are very dangerous. MacGyver would have done better by using mud, or practically anything else to plug the leak, but then there would not have been such interesting chemistry involved.

💡

Chlorine gas was first produced in 1774. What was its initial use?

The bleaching of cloth. While the dye industry had become adept at producing various colours from mineral and vegetable sources, the production of pure white cloth was a challenge. The best method available was to steep the cloth in fermented urine and then expose it to the sun. Fermented urine is rich in ammonia, which in combination with ultraviolet light from the sun can bleach cloth. But such bleaching was a dirty, smelly business. Then in 1774 along came Swedish chemist Carl Wilhelm Scheele, who discovered that a greenish yellow gas evolved when he heated a mixture of hydrochloric acid and manganese dioxide. The gas had a terrible choking effect on the lungs and quickly discoloured any leaves or flowers with which it came into contact. Scheele found that the gas dissolved in water, yielding an acid solution that today we recognize as hypochlorous acid. He didn't recognize that the gas was an element and referred to it as dephlogisticated muriatic acid. Muriatic acid was the traditional name of hydrochloric acid, and phlogiston was believed to be the mysterious component of matter that was lost upon combustion.

It took another thirty years before Humphry Davy recognized Scheele's gas as an element and named it chlorine, from the Greek *chloros* meaning "greenish yellow." But back in 1785, Claude Berthollet, the famous French chemist, had already recognized the bleaching properties of a chlorine solution. The next year the Frenchman received a visit from James Watt, of steam engine fame, and mentioned his discovery. As luck would have it, Watt's father-in-law ran a large bleaching establishment near Glasgow and was happy to hear about the discovery. Within a short time Mr. McGregor was bleaching cloth by the new process, greatly aided by his son-in-law. James Watt worked out many of the technical details of the bleaching process, including the testing of the strength of the solution by determining how much of it was needed to decolourize a standard amount of cochineal red, the classic red dye extracted from the cochineal insect. This was not James Watt's only venture into chemistry. In 1783, in a letter to Joseph Priestley, he described his determination of the composition of water. But this letter did not come to light until well after 1784, the year when Henry Cavendish independently determined the composition of water. Cavendish's discovery was immediately publicized and he gets the credit. Today, chlorine is still a widely used bleaching agent, usually in the form of calcium hypochlorite, which is made by dissolving chlorine gas in a sodium hydroxide solution. It is a lot easier than using fermented urine.

℘

The discovery of the synthesis of nylon in the 1930s
turned out to be a very profitable one for the
Quaker Oats Company. Why?

Wallace Carothers at the DuPont company discovered that two relatively simple chemicals, adipic acid and hexamethylenediamine,
could be joined together to make polymers. Industrial production
of nylon obviously relied on a cheap, readily available source of
these compounds. Oats are the base for oatmeal, or porridge as it is
also known. To convert oats into oatmeal, the outer covering of the
grain, the hull, has to be removed. This is not a useless by-product,
since the hulls contain substances called pentosans that can be converted into useful chemicals. Treatment of the hulls with acid under
appropriate conditions yields a compound called furfural, which
has a variety of industrial uses, including conversion into adipic
acid and hexamethylenediamine. Much of the furfural produced by
the Quaker Oats Company in Cedar Rapids, Iowa, went into the
production of nylon until cheaper methods using petroleum products were developed. Today, there is a trend to go back to furfural,
since it represents an application of "green chemistry," making use
of biomass instead of petroleum products.

℘

Garbage bags, shopping bags, sandwich bags,
squeeze bottles and moisture barriers in construction
are all made of polyethylene. What is the major raw
material used to make this popular polymer?

You may be surprised to learn that the major raw material for the synthesis of polyethylene is not derived from petroleum. About 70 percent of all polyethylene is made from natural gas. Natural gas is a mixture of 90 percent methane, 5 percent ethane and 5 percent propane. These components are separated and the ethane is converted to ethylene, which in turn is polymerized into polyethylene. There are two kinds of polyethylene in common use, low-density (LDPE) and high-density (HDPE). These names refer to how efficiently the long polymer chains are packed together. Production of the two varieties depends on the temperature, pressure and type of catalyst used when the individual ethylene molecules are joined together to form polyethylene. In low-density polyethylene, the chains have all sorts of branches, meaning that close packing is not possible. This kind of polyethylene is suitable for shopping bags, car covers and soft squeeze bottles. In high-density polyethylene, the chains are unbranched and can be packed together very tightly to form a hard variety. This can be used for water pipes, freezer bags and insulation around cables. Indeed, high-density polyethylene insulation made radar possible during the Second World War.

♀

Steel is sometimes coated with zinc to prevent corrosion. However, when the steel is to be protected against sea water, electroplating with a different metal works better. What metal is that?

Cadmium. Iron, which is the major component of steel, rusts when it comes into contact with oxygen and moisture. A thin layer of zinc on the surface of steel forms an effective barrier through which oxygen and moisture cannot penetrate. Even if the layer is scratched, protection is not lost. Zinc corrodes more readily than iron, and the iron will not rust until all the zinc is gone. This is why underground steel pipelines have blocks of zinc attached to them; the zinc is being used as a "sacrificial anode" to protect the steel. But zinc does not do well in the face of the most corrosive environment, the sea. That's because zinc reacts with salt to forms zinc chloride, which is soluble in water. As a result, the protective layer of zinc is slowly lost. Cadmium provides a different scenario. It too reacts with salt, but the cadmium chloride that forms is insoluble and is impervious to water. A layer of cadmium as thin as 0.05 millimetres can provide complete protection. That's why steel bolts coated with cadmium are used in areas near the ocean where exposure to sea water and sea air is a concern.

Cadmium compounds are highly toxic and are present in soil; they cannot be completely avoided. Unusually high levels in our bodies can cause all sorts of problems, including a weakening of bones and joints that can make movement difficult. This was first recognized in 1945 in Japan when a large number of people were affected by what came to be known as itai-itai, ouch ouch disease. It turned out that they had been eating rice grown on land contaminated with waste water from a zinc mine. Since cadmium often occurs together with zinc in nature, it was present in the waste to a significant extent. Even worse is the inhalation of cadmium oxide. In a classic case in Britain, workers were dismantling a construction tower and used an oxyacetylene torch to remove the bolts. Within a day the men were feeling ill, coughing and having difficulty breathing. All had to be admitted to hospital, and one eventually died. The bolts they were removing had been coated with cadmium, which under the heat of the welding torch reacted with oxygen to form the highly toxic cadmium oxide.

Why is the name of mustard gas misleading?

It really does smell like mustard, but it is not a gas. At room temperature mustard gas is actually a liquid and is spread in tiny droplets.
What is it? A chemical weapon. Mustard gas causes blisters when it
contacts the skin, damages the eyes and impairs lung function when
inhaled. It is hard to wash off because it is not soluble in water; in
fact on reaction with water it releases hydrochloric acid, which is also
toxic and corrosive. The Germans were the first to use mustard gas on
the battlefield, in 1917, mostly as an incapacitating agent. Only about
1 percent of the soldiers exposed to mustard gas died. Believe it or
not, mustard gas has a positive side. In 1943 German bombers struck
a convoy of Allied ships anchored in the harbour at Bari, Italy. One
ship was carrying 100 tons of mustard gas, which spilled into the
harbour. Within a month eighty-three of the men who had been
plucked from the water had died. Blood samples from the victims
showed a deficiency in white blood cells, which are among the most
rapidly dividing cells. This gave researchers the idea of using mustard
gas to destroy rapidly dividing cancer cells. The idea was a good one:
mustard gas can be used in the treatment of Hodgkin's disease.

Why did the exhaust of the Veggie Van smell like french fries?

Because its engine ran on biodiesel made from used restaurant oil.
In 1997 Joshua and Kaia Tickell crossed the United States in a
diesel Winnebago powered by fuel made from vegetable oil they
collected along the way. The idea was to show that vegetable oil

could be converted into biodiesel. Although Rudolph Diesel's original engine was intended to run on vegetable oil, modern diesel engines use hydrocarbons from petroleum. It is, however, possible through a simple chemical reaction to modify vegetable oil to allow it to be used in a diesel engine. In a "transesterification" reaction, the oil has to be heated with methanol and sodium hydroxide. The Tickells built a chemical reactor that they pulled behind their Winnebago, stopping along the way at restaurants to fill it up with used frying oil. They travelled over 10,000 miles (16,000 kilometres) on 100 percent biodiesel, demonstrating that alternative fuels do have practical potential. Since much of the oil had come from fast-food restaurants, it was little surprise that the exhaust smelled like french fries.

And what are you smelling when you smell french fries? A host of compounds, dominated by methanethiol. The others include 2,3-diethyl-5-methylpyrazine, trans-4,5-epoxy-2(E)-decenal and (Z)-2-nonenal. Put that into your engine and smoke it.

Most people inflate their car tires with air for free. But some pay to have the tires inflated with another gas. What gas and why?

The gas is nitrogen, and the hope is that it maintains tire pressure better. This may seem surprising because air in any case contains about 78 percent nitrogen, with the rest being mostly oxygen. So how much difference can inflating with 100 percent nitrogen make? Not much, unless you are driving a race car or piloting a jumbo jet. The difference comes in because nitrogen is less likely than oxygen to leak through the rubber. Since less gas is lost, inside pressure is better maintained. Many silly explanations have been offered for this effect, mostly by people who are not well grounded in science. They speak of nitrogen molecules being bigger or heavier than oxygen and therefore less likely to migrate through the rubber. This is pure nonsense. Nitrogen actually has a lower molecular weight than oxygen and diffuses through openings more quickly. The real explanation is that nitrogen is less likely to dissolve in rubber than oxygen and therefore less of it is lost from inside the tire.

Proper tire inflation is important for safety and gas mileage. An underinflated tire reduces mileage and an overinflated one makes less contact with the road, reducing traction. Car races can be lost or won based on tire performance. Another factor is that air contains water vapour, and the expansion and contraction of this is different from that of other gases. Without knowing the exact humidity of the air used to inflate a tire, it becomes harder to predict pressure changes. Again, this is important in racing, but not on your car. Because of the low temperatures at heights where airplanes fly, water vapour inside tires can freeze to chunks of ice, which would interfere with tire performance on landing. This is avoided by using pure nitrogen. There are also indications that a tire is less likely to burst into flame from the tremendous heat generated by friction on landing if there is no oxygen inside to aid combustion. Finally, as oxygen travels through the rubber of a tire, it degrades it more than nitrogen does. That's because, chemically speaking, oxygen is a free radical and readily reacts with other substances. Indeed, antioxidants are commonly incorporated into tires to increase longevity. These considerations, though, are irrelevant for the average driver, because under normal driving conditions inflating with pure nitrogen makes a trivial difference. Better just to check your tire pressure with a tire gauge every week.

Carbon dioxide is a colourless gas, yet sometimes it is referred to as "green." Why?

Because it can be used in chemical processes to replace more environmentally unfriendly and more toxic substances. "Green chemistry" is a burgeoning field founded on the philosophy of redesigning chemical processes to avoid hazardous substances. The dry-cleaning industry, for example, for decades relied on the solvent perchloroethylene, which is effective at removing stains but is toxic. Carbon dioxide can be converted to a liquid at high pressure and does an excellent job as a "dry cleaner" without the environmental baggage that perchloroethylene brings. It is a "green" solvent.

Our society is based on the use of chemistry. Just think of the medications, the plastics, the synthetic fibres, the cosmetics, the food additives and toilet tissues that define modern life. Producing these requires various chemical feedstocks, a selection of catalysts and a host that can be hazardous to ecosystems and sometimes even to humans. In the past, emphasis has been on developing novel useful products that were economically profitable, and pollution problems were addressed as they arose. When needed, or when forced, industry paid to clean up the mess created by inappropriate disposal of waste chemicals. Today, most companies realize that this is not the way to make businesses flourish and the drive is on to introduce "green" processes.

DuPont, for example, has run into a big problem because one of the chemicals used in producing Teflon, perfluorooctanoic acid (PFOA), persists in the environment and may represent a cancer risk. Teflon is produced by a process that requires oily substances to mix with water, and PFOA is an ideal emulsifier for this job. But if the reaction is carried out in liquid carbon dioxide, no emulsifier is needed. The same chemists who were clever enough to figure out how to make the myriad of products we use in our daily lives are also clever enough to find "greener" processes. Pfizer is already producing Viagra through an improved process that eliminates toxic solvents. Last century we experienced the green revolution in agriculture; this century we will see a green revolution in chemistry.

☙

A room in the famed Hermitage Museum in St. Petersburg is known for its spectacular green decor. What is this room called?

The malachite room, named for the mineral responsible for all that green. Malachite is an ore of copper, chemically corresponding to copper carbonate hydroxide. It forms in the ground when various copper minerals react with carbonated water or with limestone. Malachite was mined as early as 4000 BC by the Egyptians and was used in decorative construction as well as in jewellery. It was also ground up to produce a pigment commonly used in eye makeup. Huge deposits of malachite are found in the Ural Mountains, which may explain its use in many of the opulent buildings constructed in czarist Russia. The Hermitage was started in the eighteenth century to house Empress Catherine II's private art collection, but the Malachite Room was built in the 1830s and served as the official drawing room for Alexandra, wife of Nicholas I. The colour of malachite is such a beautiful green that when a novel synthetic dye was made by chemists in the nineteenth century, it was named malachite green. This has led to some confusion, because the dye does not contain any malachite and has no chemical similarity to the mineral. It can be used to dye silk, wool, leather and cotton and even has some biological properties.

Malachite green can be used to treat parasites and fungal infections in fish. In fact it was commonly used in commercial aquaculture until laboratory tests revealed that rats developed tumours when fed malachite green at a concentration of 100 parts per billion for over two years. For this reason malachite green cannot be used in fish destined for human consumption. The only fungicides allowed in Canada for food fish are formaldehyde, high salt concentration and hydrogen peroxide. The Canadian Food Inspection Agency routinely tests fish for the presence of malachite green and has on occasion found traces. In 2005, malachite green was found in salmon raised in one of British Columbia's largest fish farms. Company officials were befuddled and could not explain how the contaminant ended up in the fish because according to them malachite green had not been used for fourteen years. Perhaps, they suggested, it had been present in hatchery eggs that the company had purchased. In any case, some thirty-five thousand fish were destroyed, even though malachite green levels were only about 1.3 parts per billion, below the level of 2 parts per billion deemed safe in Europe. The chance of anyone's being adversely affected by eating the tainted salmon was essentially zero. And remember that you don't have to be afraid of your malachite jewellery—it has nothing to do with malachite green.

☙

To what industry has the African civet cat made a significant contribution?

Perfumery. Civet cats have a gland between the anus and genitals that produces a pungent substance to mark territory. Both males and females produce the secretion, but the male's is more potent. One of the main compounds in this secretion is civetone, which has an unpleasant smell when concentrated but becomes appealing when diluted. This odoriferous substance has long been an item of commerce, albeit not a cheap one. The civet cat does not give up its chemical supply readily. The traditional method of collecting the scent is as unpleasant as the raw scent itself. The male civet cat is placed in a cage after capture from the wild and is taunted with a stick waved in front of its face. When it grabs the stick, the back of the cage is opened and the oil is removed from the scent sac in the genital area. The rear legs are held as the secretion is collected with a spatula. Undoubtedly the animal does not enjoy the performance. Each cat can produce about 25 to 30 grams of the piercing sweaty aroma a month, but in perfumery that goes a long way because so little is needed. The extract initially smells so bad that it can be adulterated with human baby feces to increase its weight.

Civet cats, and musk deer as well, undoubtedly consider Wallace Carothers a hero. The chemist who is famous for inventing nylon was also the first to produce synthetic civet-like and musk-like odours in the laboratory. Today, almost all the civetone used by the perfume industry is produced synthetically. So civet cats no longer have to worry about being tormented to give up their perineal gland secretions. It is people who now worry about being exposed to fragrances that may trigger asthma, skin reactions and chemical sensitivities.

What is the purpose of a cosmetic product that contains a tyrosinase inhibitor?

To lighten "age spots" on the skin. One of the signs of aging is the development of brown spots on the skin, which are sometimes referred to as liver spots. They actually do not have anything to do with liver malfunction, but there is some justification for the use of the term. The liver produces an enzyme known as tyrosinase that converts tyrosine, an amino acid in the diet, to a compound commonly known as dopa, which in turn forms dopamine, an important neurotransmitter. Dopa is also used by the pigment-producing cells in the skin to make brown melanin, but this does not originate in the liver. Melanocytes, the cells that produce pigment, also make tyrosinase and use it to make melanin from tyrosine. As we age, melanocytes congregate in some areas of the skin and produce highly pigmented areas, which may be considered unsightly. One way to battle this problem is to use substances that inhibit the activity of tyrosine.

The classic substance used was lemon juice, which contains vitamin C and citric acid, both of which inhibit the enzyme. This is also the reason why lemon juice can prevent cut apples from darkening; it is the tyrosinase released from damaged cells that causes the browning effect. Hydroquinone, a synthetic compound, is more effective than lemon juice because it not only inhibits tyrosinase but also destroys melanin. Unfortunately it can also destroy skin, so its dosage has to be carefully controlled. Because this is difficult, producers have looked for alternatives. Kojic acid, isolated from a fungus, appears to be safe and effective. It is also readily available because it is produced commercially to be converted into maltol, a substance used to give breads and pastries a "freshly baked" odour. Mulberry extract also is an effective tyrosinase inhibitor and is found in a number of skin-lightening products, sometimes along with licorice and scutellaria extracts. Also commonly incorporated are alpha hydroxy acids, which remove surface skin so that the tyrosinase inhibitors can reach the melanin-producing cells.

♀

A device exchanges calcium ions for sodium ions. What is its purpose?

To soften water. Water is a great solvent, and as it flows through the ground it dissolves a number of naturally occurring minerals, including ones that contain calcium and magnesium. Rainwater is slightly acidic because it interacts with carbon dioxide in the atmosphere to form carbonic acid. This acidity increases the dissolving power of water by allowing it to react with insoluble substances to form soluble ones. Calcium carbonate is a typical case. It occurs naturally as chalk, limestone and marble, all of which are insoluble in water. But when acidic water flows over such deposits, calcium carbonate goes into solution as calcium hydrogen carbonate. This calcium-laced water is what we know as "hard water." It presents several problems. When such water is heated, the soluble calcium hydrogen carbonate breaks down and yields insoluble calcium carbonate. This is essentially the reverse of the formation of hard water, and it is what causes scale formation in kettles and hot water pipes.

Hard water presents yet another problem. Dissolved calcium reacts with soap to form a precipitate, which we recognize as the "bathtub ring." It also leaves laundry looking grey. One way to reduce water hardness is to run the water through an ion-exchange filter, which replaces calcium ions with sodium ions. Sodium in the water does not have the same problems associated with it as calcium because it does not form insoluble compounds. Another method is to add washing soda to the water. This is sodium carbonate, which reacts with dissolved calcium to form insoluble calcium carbonate, thus eliminating calcium from the water.

Another type of hardness, known as permanent hardness, occurs when water dissolves calcium sulphate, or gypsum, from the ground. This reaction is not reversible with heat, but the calcium can still be removed with washing soda or with an ion-exchange filter. As far as drinking water goes, hard water appears to be healthier. Studies have shown a lower rate of heart disease in hard water areas, probably thanks to the beneficial effects of increased calcium and magnesium intake.

♀

Where in a home would you find radioactive americium?

In a smoke detector. Fires kill more people in North America than all natural disasters combined. Many people die when they are suddenly overcome by smoke without any warning. Smoke detectors can provide the warning and allow the time needed to escape. There are two types of smoke detectors. Photoelectric detectors work on the principle that smoke particles can scatter light. Just think of how the path of a sunbeam shining into a darkened room is made visible by light reflected from dust particles. In a photoelectric detector a tiny light beam from a light-emitting diode shines across a little chamber. A light detector is located in another part of the chamber, out of the path of the light beam. But when smoke particles are present, they scatter the light towards the detector. The light hitting the detector triggers an electrical circuit that activates the alarm. The more popular type of smoke detector, known as an ion-chamber or ionization detector, employs a small amount of a radioactive material to generate an electric current. Americium-241 (half life 458 years) is made by bombarding plutonium-239 with neutrons. Am-241 radiates electrically charged particles known as alpha particles, which bombard air molecules inside the detector, knocking off electrons and generating ions. These ions complete an electrical circuit. The presence of smoke particles or gases reduces the mobility of these ions and thus reduces the electric current. The reduction in current sets off the alarm.

Ion-chamber detectors are useful because they can be set off by gaseous combustion products that form even before there is any smoke. This also means that the detectors can be triggered by innocuous substances such as steam from a shower—a minor annoyance, but nothing compared with the benefits offered. The benefits, of course, are possible only if the detector is working properly. Pushing the button only tests the alarm's circuitry. The detector should be tested every couple of weeks by blowing out a candle just under it. Ionization detectors have a lifetime of around ten years, because the americium source does wear out. Some peo-ple are concerned about the fact that these detectors use a radioac-tive substance. The amount of americium used is very small and is well shielded. Furthermore, alpha particles travel only about 2 inches (5 centimetres) from their source, so the only way americium would present a danger would be if the source were eaten. So don't eat smoke detectors. And above all, remember that even the best smoke detector is useless if the battery doesn't work. Statistics tell us that about 25 percent of detectors are non-functional, mostly because the battery has worn out or has been removed to power a toy. If you do that, you could be playing with your life.

brain fuel

♀

If you were using cathodic protection, what would you be trying to protect?

You would be trying to prevent iron from rusting. The rusting of iron is an expensive process. It is estimated that the deterioration of iron due to corrosion costs billions of dollars a year. The chemical process is quite simple. Iron reacts with oxygen from the air to form iron oxide. This is termed an electrochemical reaction because the oxygen actually steals electrons from the iron. Water is required for this reaction to proceed, and the process is faster if the water contains substances called electrolytes, which can carry an electric current dissolved in it. Salt is a great electrolyte.

A process known as cathodic protection can be used to prevent rust formation. The iron to be protected is attached to another metal such as zinc or magnesium, which give up electrons to oxygen more readily than does iron. The so-called sacrificial cathode will then corrode and the iron will not. Underground gasoline storage or oil tanks can be protected in this fashion. This principle is readily demonstrated with a simple experiment. Take two nails, attach a piece of zinc to one of the nails and immerse both in salt water. Watch the difference in corrosion! Rusting can also be prevented by excluding oxygen and moisture. Paint does this quite well. Another possibility is to alloy iron with other metals such as chromium to make stainless steel. In this case chromium reacts with oxygen to form chromium oxide that deposits as a thin impermeable layer on the surface of the metal and protects the iron underneath. Iron can also be coated with a thin layer of another metal that is less prone to oxidation. So called "tin" cans actually are made of iron coated with a thin layer of tin.

♀

Researchers are looking at using solar energy to convert zinc oxide to zinc. If an efficient method can be found, it could lead to a reduced dependency on gasoline for cars. What would the zinc be used for?

The production of hydrogen. Cars can be made to run on hydrogen, which is a clean-burning fuel and does not produce carbon dioxide when combusted. The problem is the availability of hydrogen. It can be produced by electrolysis, a relatively simple procedure that should be familiar to all high school students. With the passage of an electric current, water molecules break down to yield oxygen and hydrogen. The problem is that this process requires a significant amount of electricity, making it economically not viable. But in another reaction that should also be familiar to high school students, water reacts with metallic zinc to produce zinc hydroxide and hydrogen gas. Here the problem on an industrial scale is the production of pure zinc in large quantities. Zinc oxide is abundant in nature and can be converted to zinc by heat, but it takes a great deal of heat, which has to be furnished somehow. Using fossil fuels to heat the zinc is not the answer—it obviously defeats the purpose of producing zinc to make hydrogen. That's why researchers are looking at heating zinc oxide using solar energy.

The largest solar research facility in the world is at the Weizmann Institute in Israel, where a large array of computer-guided reflective mirrors follows the sun and focuses the sunlight to produce tremendous heat in a solar furnace. When zinc oxide is placed in the surface along with some coal, the carbon in the coal strips away the oxygen to leave zinc metal behind. The carbon monoxide by-product can be easily dealt with. The zinc that is produced is

ground to a fine powder and can be easily and safely transported.
Hydrogen can then be produced wherever needed by combining the
zinc powder with steam. Zinc oxide is the other product of this
reaction and can be recycled into zinc using solar energy.

𝄡

Nitric and nitrous acids are major contributors to acid rain. What is the most common natural source of these acids?

Lightning. These acids form when oxides of nitrogen dissolve in
rainwater. The oxides in turn form when the energy of a lightning
bolt allows nitrogen and oxygen, the natural components of air, to
combine. Nitrogen makes up about 78 percent of the atmosphere,
oxygen about 20 percent. These gases happily coexist without reacting with each other, which of course is fortunate for us since we
need oxygen to survive. But if nitrogen and oxygen are heated to a
high enough temperature, they do react. First they combine to form
nitric oxide, or NO, which reacts further with oxygen to form
nitrogen dioxide, or NO_2. This gas in turn dissolves in water to
form nitric and nitrous acids. Lightning is the most significant
source of naturally occurring oxides of nitrogen, but man is a
greater contributor than nature. Coal- or oil-fuelled electrical
power plants, cars, buses and airplanes generate far more oxides of
nitrogen than lightning does.

   While nitric and nitrous acids are a problem, sulphuric acid is a
greater contributor to acid rain. Its source is elemental sulphur,
which is found in coal and oil in small amounts. When it burns it
yields sulphur dioxide, which in turn reacts with oxygen to yield

sulphur trioxide, which dissolves in rain to form sulphuric acid. Volcanoes and sea spray also contribute some sulphuric acid, but far less than human activities. Acid rain can harm aquatic life, interfere with plant growth and eat away at buildings. Sulphate particles in the air can even interfere with visibility. Technologies exist to remove nitrogen oxides and sulphur dioxide from industrial emissions, and laws are in place to force industry to make use of these methods. Of course, there is nothing we can do about lightning.

A Canadian cosmetics company advertises its PlantLove Botanical Lipstick as having a biodegradable tube made of polylactic acid. What is the raw material used to make polylactic acid?

Corn. Everyone is trying to be eco-friendly these days, with terms like "biodegradable" and "renewable resource" being plastered all over consumer products. "Biodegradable" refers to substances made of organic compounds that under the right conditions are broken down by microbial organisms into simple, safe substances such as carbon dioxide, methane and water. Made from a "renewable resource" generally means that the raw materials used to make a product derive from a plant source rather than from petroleum or natural gas. Polylactic acid is a plastic that meets the biodegradable and renewable resource criteria, but there are some buts here in terms of "greenness." First of all, the plastic does indeed biodegrade in under sixty days, as long as it winds up in commercial composting installations. The likelihood of a lipstick container ending up in one of these, however, is small. It is much more likely to end up in a landfill, where polylactic acid does not biodegrade. What about corn being a renewable resource? Of course that is true. But it is also true that growing corn requires fertilizer and pesticide input, and the environmental consequences of the transport systems required are not inconsequential. Lactic acid is made from corn through a bacterial fermentation process, and one ear of corn can yield enough plastic to make twelve lipstick tubes. But converting lactic acid into polylactic acid requires a sophisticated process employing various tin catalysts and is not a totally environmentally benign endeavour.

Now what about the lipstick inside the tube? The basic ingredients, such as castor seed oil, meadowfoam seed oil, beeswax and carnauba wax, are plant derived, so can be classified as renewable, but the product uses the same array of colourants and preservatives as other lipsticks. Butylated hydroxytoluene (BHT) and butylparaben are effective preservatives and are commonly used in numerous cosmetic products. PlantLove Botanical Lipstick, made by Cargo, also includes a trademark "orchid complex" alleged to fight free radicals, reduce fine lines and boost immunity. This is sheer marketing—there is no evidence to back up any such claim. Admittedly, the product does come in an interesting package. The box is biodegradable and is embedded with wildflower seeds. You just plant the whole box and soon you can admire a bouquet of flowers. But you'll have to shell out lots of green for this green product. A tube of Botanical Lipstick goes for twenty dollars. Cargo is donating two dollars from every sale to St. Jude Children's Research Hospital, surely a noble gesture. Of course, you could always buy a tube of comparable lipstick for a few dollars and donate the difference between that and twenty to the hospital. A better deal all around. But you won't be getting colours designed by the likes of Lindsay Lohan and Mariska Hargitay. Oh well.

♀

What environmental problem is referred to in Ireland as "witches' knickers"?

Plastic bags that have been carelessly discarded and get caught up on tree branches. Polyethylene bags fluttering in the wind, desecrating our beaches and piling up in cabinets under our sinks, waiting for uses that never seem to materialize, have become symbolic of our throw-away society and lack of environmental conscience. We use a lot of plastic bags, that much is for sure—worldwide, roughly a million every minute of every day. Why? Because they are convenient to use and cheap to produce. And what happens to them? Some of course are reused as garbage bags or as pooper scoopers, and a few people even reuse them for grocery shopping. But most are discarded and end up in landfills. Since the bags are highly compressible, they actually make up only about 1 percent of landfill waste. Bags in a landfill are not a problem, but those that escape into the environment are. Although they make up only about 2 percent of all litter, plastic bags can potentially harm wildlife and even people. Whales and turtles can mistake plastic bags for jellyfish and eat them, often with fatal results.

In Bangladesh, where garbage cans are virtually non-existent and waste collection is poor, plastic bags were routinely dropped in the streets, ending up being washed into rivers and sewers where they clogged dams and drainage systems, causing flooding. In 2002 Bangladesh took the drastic step of banning all plastic bags and fin-ing anyone caught using one. In the northern Indian state of Himachal Pradesh you can be jailed for seven years for using a plas-tic bag. Ireland has had success with a tax on all bags, and many cities in North America are considering ways of reducing plastic bag use, either with outright bans or by mounting effective recycling programs. Technically, polyethylene bags can be recycled into various useful items, ranging from plastic lumber to other bags, but it is a question of economic viability. California has introduced a law making it compulsory for stores above a certain size to offer recycling programs. Biodegradable shopping bags made of starch are being widely promoted, but there are questions about the conditions under which these will truly biodegrade.

Plastic bags are not the real problem. We are. People have become too accustomed to a disposable society. Habits need to be changed. Taking a reusable cloth bag to the supermarket is a great idea, but realistically, this is not going to happen to any significant degree. Paper bags are not the answer, as they also produce litter and take up much more space in landfills. Using fewer bags, perhaps motivated by a charge for them, is part of the answer, but it is in recycling that real potential lies. Right now only about 1 percent of bags are recycled, and that needs to be dramatically increased. Bags can also be burned as a source of energy. Instead of looking at polyethylene bags as an environmental scourge, we need to find ways to make them into a viable commodity after they have served the consumer in a useful fashion.

♀

A chicken bone immersed in vinegar for a few days becomes as flexible as if it were made of rubber. What has the vinegar removed and what has it left behind?

Calcium has been removed and elastin and collagen, the proteins on which calcium is deposited, remain. Contrary to what many people think, bone is not static but dynamic. It is constantly being reshaped by osteoblasts, cells that build bone, and osteoclasts, cells that break down, or "resorb," bone. Bones are complex structures, but essentially consist of a framework of connective tissue composed of the proteins collagen and elastin, which act as support for calcium phosphate, the mineral component of bone. The osteoblasts that crank out collagen and coat it with calcium eventually wind down their activity and become osteocytes, cells that become entrapped in the bone matrix and lend support to the structure. Osteoclasts can break down any extra bone and can also supply calcium to the bloodstream from deposits in the bone in case there is a shortage of supply. Calcium is needed by muscle and nerve cells for proper function, and bones serve as a repository. Acetic acid in vinegar reacts with calcium phosphate in bones and converts it into calcium acetate, which is water soluble. In this fashion calcium is leached out from the bone, leaving behind the flexible elastin and collagen proteins. Gives a whole new meaning to the expression "rubber chicken." If collagen is boiled in water, the long chains of amino acids break down into shorter ones to yield useful substances such as gelatin and glue.

❦

Hydrogen is looked upon as the fuel of the future for powering automobiles. Prototype hydrogen cars already exist. Where does the hydrogen come from?

The combustion of natural gas. While there are many ways of producing hydrogen, most today comes from the burning of methane, better known as natural gas. When methane is combined with steam at a high temperature, carbon monoxide and hydrogen gas are produced. The carbon monoxide can be further reacted with steam to yield more hydrogen and carbon dioxide. While this process does produce carbon dioxide emissions, just like burning gasoline in an automobile, the emissions occur at a single location and the carbon dioxide can be captured. The hydrogen produced can power a car in two ways. It can be directly burned in the engine, just like gasoline, yielding only water as a product. Or the hydrogen can be used in a fuel cell, where it combines with oxygen to produce electricity, which then runs an electric motor. This is essentially the reverse of the classic electrolysis reaction in which an electric current is passed through water to yield hydrogen and oxygen.

There are various problems with a "hydrogen economy." Hydrogen is really only a means of storing energy, because it takes energy to produce it. Right now hydrogen cannot be produced economically enough to replace gasoline. Inventive ways making use of wind power, solar energy or nuclear energy will have to be found in order to use hydrogen efficiently. There is also the problem of handling hydrogen gas, which will have to be stored in vehicles under high pressure.

The fuel cell effect was discovered by the Swiss-German chemist Christian Friedrich Schönbein, with the first description in English appearing in 1839 in the *Philosophical Magazine*. Based on this idea, Sir William Robert Grove, a London lawyer who dabbled in engineering, built a functioning fuel cell. Interestingly, during his experiments with electricity Schönbein noted a strange smell in the air, which he later identified as ozone. Ironically, today one of the plusses of fuel cells is that they reduce ozone pollution associated with cars that run on gasoline.

Schönbein is perhaps best known for his invention of gun cotton, which he supposedly discovered after wiping up a mess of nitric and sulphuric acids on the floor with his wife's cotton apron.

When he hung the apron in front of the fireplace to dry, it went up in flames but produced no smoke. The era of smokeless gunpowder was under way. Schönbein licensed his discovery to an English entrepreneur, but production ended in tragedy when an explosion took the lives of two dozen workers. After this Schönbein discontinued work on gun cotton, but his ideas were later used by Alfred Nobel to develop dynamite.

just amazing

Australians have been warned that jars of a commercially available fertilizer made from a liquefied animal product may spontaneously explode. What animal was used?

The cane toad. When Australia began to cultivate sugar cane back in the 1930s, farmers ran into a problem: the cane beetle. The larvae feast on the roots of the sugar cane and either kill or stunt the growth of the plants. In Hawaii, where lots of sugar cane was grown, the beetles did not have much of an impact. That's because the native cane toads found the beetles to be tasty morsels and controlled their population. Based on the Hawaiian experience, Australia decided to import about a hundred toads, hoping they would multiply and gorge themselves on beetles. A little "love pond" was even set up to facilitate the toads' romantic escapades. It worked. Cane toad eggs began to hatch with an impressive frequency, and in a short time some three thousand adults had been marshalled for release into the sugar cane plantations of Queensland. And then the toads became a pest themselves. They reproduced at an amazing rate and soon were fearlessly hopping all over the countryside. Why fearlessly? Because cane toads have an astonishing protective mechanism.

The toads produce a poison called bufotenin in the parotid glands on the back of their necks that effectively deters predators. Dogs can die if they bite into one of these toads. An irate toad can even spray its venom at an enemy; the poison can be absorbed through mucous membranes such as eyes, mouth and nose and in humans may cause intense pain, temporary blindness and inflammation. People have even died after eating toads or soup made from boiled toad eggs. Others have ended up in hospital after licking a toad in a misguided effort to get high on the "toad juice." The toads have become such a pest in Australia that various methods have been devised to control them, including publicity campaigns to catch the critters and do away with them. But what to do with the corpses? Well, a company has come up with an idea. Liquefy the remains and market it as a fertilizer. Unfortunately the inventors did not consider that the toad juice might ferment in the bottles and produce a gas, which in turn could cause an explosion spewing the sticky, foul-smelling contents all over the place. Their advice? Loosen the caps on unused bottles of ToadJus before storing them.

Incidentally, the cane toads never did solve the cane beetle problem in Australia. It seems there were just too many other goodies to gobble up in the Australian countryside.

What is the military use of Silly String?

To detect trip wires linked to explosives. Silly String is a party item loved by children and hated by cleaning crews and environmentalists.

It emerges on command from spray cans to form long, colourful foamy strings. The foam is so light that it can settle on trip wires without triggering an explosion. Soldiers in Iraq have made great use of Silly String to detect booby traps. Before going into a house, they spray the goop all over, and if the strings stay suspended in air, a blast is in the offing. Silly String is an ingenious chemical concoction, but its exact composition is difficult to pin down. Manufacturers guard the secret closely to protect their profits. But the general concept is clear. The original patent describes a solution of polyisobutylmethacrylate in Freon with sorbitan trioleate as a foaming agent. Freon, a gas under ordinary conditions, can be liquefied under pressure.

In the case of Silly String, Freon acts as both a solvent and a propellant. Depressing the button on top of the can opens a passageway through which the contents of the can can emerge. The greater pressure inside the can now forces the solvent and everything dissolved in it to spray out. As the solvent quickly evaporates, it leaves behind the plasticky residue. Since evaporation is quickest from the surface of the string, this part hardens the fastest. Some of the propellant gas is then trapped inside the string and creates the holes characteristic of a foam. The original type of Freon used has been banned because of its potential destructive effect on the ozone layer. Today, manufacturers use either newly developed Freons that are not implicated in ozone destruction or petroleum-based hydrocarbon solvents. Unfortunately, the hydrocarbons have the problem of flammability. Some municipalities have banned Silly String because it makes a mess and can clog sewer systems, but in Iraq, where lives are at stake, it is a welcome addition to a soldier's equipment.

What substance named after the Greek word for "smell" was once pumped into the London Underground because it was believed to be invigorating but today is considered to be an air pollutant?

Ozone, named from the Greek *ozein* for "smell." Ozone is a bluish gas produced when oxygen is exposed to a high voltage. It has a characteristic fragrance commonly noted around photocopy machines and high voltage wires and during thunderstorms. Victorians believed ozone to be health giving and invigorating, and it was pumped into hospitals, churches and the underground. Today we realize ozone can impair lung function. Cars emit various oxides of nitrogen that, with the aid of sunlight, can convert oxygen in the air to ozone. Curiously, the same gas in the stratosphere is beneficial; it protects us from excessive exposure to ultraviolet light.

Nylon played an important role in the movie *The Wizard of Oz*. What was that role?

As we all know, Dorothy was scooped up by a tornado and carried to the land of Oz. Well, the tornado wasn't really a tornado. It was a nylon stocking, filled with earth, blown by a fan and manipulated from above in a swirling motion. Nylon had been invented by Wallace Hume Carothers at DuPont in 1935 and was introduced to the world at the 1939 New York World's Fair. The appearance of the nylon stocking in the film actually came before women had a chance to try the newfangled stockings fashioned out of petroleum derivatives. Nylon stockings were first sold in the United States in

1941, but there was a great shortage because American servicemen were stockpiling them to give as gifts to British ladies. Indeed, in Britain nylon hose became the most asked-for gift from American servicemen. Many a war bride was wooed with nylon.

♀

Why have some convenience stores installed blue lights in their washrooms?

To prevent addicts from using the washrooms to inject their drugs. The veins on the back of the hand are a common injection site and show up as blue lines under the skin. Under a blue light, though, everything looks blue and the veins essentially disappear from sight. The hope is that addicts will then stop using these facilities to further their habit. Convenience stores are not the only ones to have tried this technique to frustrate addicts. Some hotels and bars also have installed blue lights in their washrooms, and in Australia, Britain and parts of Europe, hospitals, parks, universities and even department stores have taken this route. Although some bar owners claim that the needle-collection bins in their washrooms fill up more slowly, health authorities are not convinced of the effectiveness of the blue light scheme. Some believe that making injection more difficult just increases the risk of injury and overdosing.

In any case, experienced users can find distended veins even in the dark, or they can simply mark the injection site with a pencil before going into a blue-lit washroom. But the attempt to use such lights in public washrooms does demonstrate the extent of the drug problem in our society. In Regina, Saskatchewan, for example, there are an estimated three thousand injection drug users, and authorities hand out some 1.4 million needles annually to try to decrease injection-related diseases such as hepatitis and AIDS.

☙

The April 1907 issue of *American Medicine* featured a paper by Dr. Duncan MacDougall describing his experiment whereby the beds of dying patients were placed on a sensitive balance. What was Dr. MacDougall trying to measure?

The weight of the human soul. The paper was titled "Hypothesis Concerning Soul Substance Together with Experimental Evidence of The Existence of Such Substance." MacDougall, of Haverhill, Massachusetts, placed six dying patients on the specially constructed balance and concluded that at the moment of death there was a loss in weight of about three-quarters of an ounce, or 21 grams. He had previously determined the weight loss attributed to evaporation of moisture from the skin, and by comparison this loss was sudden and much larger. He even accounted for weight loss due to urine and fecal eliminations and concluded that these could not account for the change in weight. Air loss from the lungs was not the answer either, as he determined by lying on the scale himself and noting that breathing had no effect on weight. After weighing his six patients, MacDougall went to work on dogs. How he got his hands on fifteen dying dogs is not clear, but he found no weight loss at the moment they expired. He wasn't surprised, of course, because he didn't think dogs had souls. No one since has confirmed MacDougall's findings, but the 2003 movie *21 Grams* was based on his idea.

Roughly from the sixteenth to the eighteenth century people would sometimes drink wine that had been left standing overnight in a cup made of antimony. Why?

They were trying to expel bad "humours" from the body. One way to expel such humours, which were thought to be the cause of illness, was by inducing vomiting. And compounds of antimony are very adept at doing that. Wine always contains some tartaric acid, which can react with the elemental antimony in the cup to form antimony tartrate, a potent emetic. While small amounts of antimony—50 milligrams or so—induce vomiting, not much larger amounts induce death! Playing around with antimony "cures" was a dangerous business indeed, and one of its victims may have been Wolfgang Amadeus Mozart. Nobody knows for sure what killed the thirty-five-year-old composer in 1791, but the fever, vomiting and swollen belly he suffered from were characteristic of antimony poisoning. Mozart had always been sickly, and it is well known that he had often been treated with antimony compounds by his physicians and that he even dosed himself when he didn't feel well.

Mozart actually believed he was being poisoned, but not by himself. He thought his musical rival Antonio Salieri was trying to do him in. Although this possibility is alluded to in *Amadeus*, Peter Shaffer's play and its movie adaptation, historical facts do not corroborate the poisoning story. Contrary to the portrayal, Salieri did not confess at the end of his life to having tried to kill Mozart. Of course, we cannot prove that antimony was responsible. Mozart had also suffered from rheumatic fever since childhood, and this may have led to his ultimate demise at a young age.

Today, antimony is no longer used medically, but its toxicity still sometimes makes the news. A volatile compound of antimony known as stibine (SbH3) was suspected of being responsible for crib death in the 1990s. The theory was that it was produced from antimony oxide added as a flame retardant to polyvinylchloride sheets. A fungus found in mattresses supposedly made this conversion possible, at least under laboratory conditions. The theory has now been dismissed because neither the fungus nor levels of antimony in babies' blood could be correlated with crib death.

☿

Natives in Samoa wrap the grated seeds of the futu tree in hibiscus leaves and toss the packets into tidal pools. Why?

To stun the fish, which then rise to the surface and can be collected by hand. The futu tree (*Barringtonia asiatica*), also known as the fish poison tree, grows in coastal areas of the tropics and can reach a height of some 65 feet (20 metres). Compounds present in its seeds have piscicidal, or fish-killing activity. One of these, known as ranuncoside, has been isolated and its molecular structure has been identified. The seeds of the tree are usually scraped on lava rocks or on coral and are then wrapped in leaves of the wild hibiscus tree. Holes are poked in the leaves before the packet is tossed into the water to stupefy the fish. The poison does not affect humans who eat the fish.

The futu tree is not unique; over a hundred piscicidal plants have been identified around the world. Some of the active compounds from such plants have been isolated and used in the management of unwanted species in waterways, in fish culture operations and in sport fishing. The most widely used such compound commercially is rotenone, which can be used to kill undesirable species in reservoirs. It decays quickly so does not present a risk to people who eventually drink the water. There is interest in such compounds because some have already been shown to have medicinal properties such as anti-ulcer and anti-tumour effects and even aphrodisiac properties. Even the futu tree may have a medical use. In some West African countries and in Polynesia, a liquid from the crushed bark of *Barringtonia asiatica* is used to treat chest pain and heart problems.

In 1940 Professor Frank Cotton, a physiologist at the University of Sydney, used women's rubber bathing suits to solve a problem faced by fighter pilots engaged in aerial combat. What was the problem and how did he solve it?

Cotton invented the G-suit and thereby solved the problem of pilots passing out when they made sharp turns with their airplanes. As jet aircraft were being developed during the Second World War, pilots faced a new difficulty. At such great speeds, as the plane banked, blood rushed from a pilot's head into his body. This could cause temporary blindness, and sometimes the pilot would even pass out. Not a good thing for a fighter pilot. Cotton thought that an inflatable suit could apply pressure to the body and force the blood back into the head. He fashioned a prototype out of two rubber bathing suits, and by 1944 Allied airmen were wearing the G-suits that we now commonly associate with fighter pilots and, of course, astronauts.

❦

Natives in the Amazon grow a plant for food but also use it for hunting. What is this plant?

Bitter cassava. The root of the cassava plant produces tubers, much like the potato. It is a dietary staple in many areas of Africa and South America. But there is a catch. Some varieties, like the bitter cassava, produce tubers that are toxic unless they are properly prepared. And we are not talking about mild toxicity, we are talking about hydrogen cyanide, a potent poison. Cassava contains a compound called linamarin, which liberates cyanide when it comes into contact with the enzyme linamarase. This enzyme is released when the cells of cassava roots are ruptured. Such rupture can take place during digestion, releasing cyanide into the blood. Over the centuries natives learned how to avoid such poisoning by macerating the tuber, thereby releasing the cyanide, much of which evaporates. The rest is removed by repeatedly washing the mashed cassava with water. Amazon hunters realized that the cassava's toxin could be adapted to hunting. They used the sap from the root of the bitter cassava on the tips of blow darts to quickly bring down prey.

❦

What would cause pink flamingos in captivity to lose their colour?

A diet different from what the birds would consume in the wild. The pink of flamingos comes from naturally occurring pigments called carotenoids that are found in the algae and small shrimp the birds consume in the wild. Without a supply of carotenoids in their food, flamingos will lose their colour. Salmon also get their colour from carotenoids, specifically from astaxanthin, which is abundant in the krill they eat. Farmed salmon, which do not have access to carotenoid-containing algae or krill, have their diet supplemented with astaxanthin, which can be produced synthetically or isolated from commercially farmed algae. Astaxanthin actually occurs in nature in several closely related forms known as stereoisomers. The ratio of these isomers to each other is somewhat different in krill, in farmed algae and in the synthetic version of astaxanthin. This has no effect on colour but it does allow chemists to determine the source of the pigment.

Wild salmon these days has greater commercial appeal than the farmed variety because of publicity given to studies that have suggested a higher level of contaminants such as PCBs in farmed fish. Although the differences are minor and unlikely to have an impact on health, consumers are willing to pay a higher price for what they perceive to be a safer product. Unfortunately this has also resulted in fraud. In some cases farmed salmon is being sold as wild salmon. We know this thanks to an investigative report published in *The New York Times*. Reporters bought samples of "wild salmon" from a number of outlets and sent these for astaxanthin analysis. Some of the fish were found to be farmed. Although this is not a safety issue, consumers should be getting what they pay for.

Sometimes a related carotenoid, canthaxanthin, is also added to fish feed. The amount that can be added is carefully regulated because the compound has been linked to an eye problem called canthaxanthin retinopathy. This disorder occurs when canthaxanthin or its metabolites crystallize around or on the retina, blocking nerve signals and causing white flashes and other visual problems. That's one of the reasons why tanning pills containing canthaxanthin have been banned. Here the idea was to achieve a tan without risking injury from ultraviolet light by colouring the skin from the inside. Carotenoids are fat soluble and can build up in the fatty layer under the skin, giving the illusion of a tan. The colour, however, was not always great, and sometimes the "tanner" ended up looking like a pink flamingo.

What is hardware disease?

No, it isn't rust. It is something that happens to cows when they consume bits of metal. You may think that cows dine only on grain or grass, but that isn't exactly so. Nails, staples and bits of baling wire can end up in the feed and cause real problems. The digestive tract gets irritated, resulting in inflammation, which can reduce weight gain and milk output. The cow can also get depressed and lethargic. How do you prevent this from happening? By feeding some metal to the cow. Not ordinary metal, but a magnet. So-called cow magnets are available for this purpose. The animal is fed the magnet, which lodges in its forestomach and collects scraps of metal, preventing further penetration. A single magnet lasts the lifetime of a cow. Hardware disease isn't limited to bovines. There is a literature reference to hardware disease diagnosed in a goose, but it isn't clear just what sort of metal the goose was feasting on.

A Washington State University scientist has invented a product made of carnauba wax that is to be sprayed on apple trees. What is the purpose of this spray?

To protect the apples from sunburn. Professor Larry Schrader estimates that as much as 10 percent of Washington state's apple crop, a value of some $100 million, is lost every year to sunburn. When apples get hot under the sun, exposed cells die and black spots appear. It doesn't take long for this to happen. If the surface temperature of an apple reaches 125°F (52°C), which can easily happen in the sunshine, it takes only ten minutes for damage to occur. The common technique to prevent this is to cool the apples with a spray of water, but this of course is a problem in times of drought. That is where Schrader's invention, Raynox, comes in. The waxy liquid, which is made from the wax found on the leaves of carnauba palm trees in Brazil, is a natural blocker of ultraviolet light. When sprayed on apple trees, it can reduce sunburn by an average of 50 percent. The patented product is now used on about 30,000 acres (12,000 hectares) of the state's 200,000 acres (81,000 hectares) of apples.

In 1978 the mayor of San Francisco and a city supervisor were murdered by a disgruntled former employee. The defence argued that the assailant's addiction to which junk food was responsible for the crime?

Twinkies. Dan White, a former city supervisor, shot mayor George Moscone and supervisor Harvey Milk. The defence argued that

White had been depressed, which led to his eating more and more junk food, which in turn depressed him further because he knew it was not good for him. Depression was used in a plea for diminished capacity, suggesting that White was incapable of premeditated murder. The jury bought this "Twinkie Defence" to some extent and brought in a verdict of voluntary manslaughter; White was sentenced to only seven years and eight months in prison.

James Dewar invented Twinkies in 1930 and came up with the name after seeing a billboard for Twinkle Toe Shoes. Originally the filling was banana cream, but when there was a shortage of bananas during World War II, vanilla cream was substituted. Today, Americans eat 500 million Twinkies every year, filling themselves with a filling of sweetened shortening. The amount of cellophane used to wrap these would encircle the globe one and a half times.

Rumours that Twinkies will last forever are unfounded. The actual shelf life is seven days. Twinkies will not cause criminal behaviour, unless one considers the formation of deposits in coronary arteries to fall into this category. This collage of fat, with plenty of trans fatty acids, and sugar is a nutritionist's nightmare. But you can still have fun with Twinkies without eating them. The Society of Physics Students at the University of Washington sponsors the annual Twinkie toss. They have tried to propel Twinkies to great distances with the likes of surgical tubing and golf clubs. They are also conspiring to get Twinkies into space, which is just where they may belong. And some-one has even come up with a Twinkie milkshake. Just mix 3 Twinkies, 2 scoops of ice cream and 1 cup of sugar in a blender. Process till smooth and mix in a shot of rum or Wild Turkey bourbon.

And what happened to Dan White? He was released from prison after five years and then committed suicide. James Dewar, on the other hand, died at age eighty-eight and attributed his longevity to eating two Twinkies every day. One wonders how long he would have lived if he *hadn't* indulged.

♔

Why should dogs beware of the Grotto del Cane?

A dog that ventures into this cave near Naples, Italy, will be suffocated by carbon dioxide gas. The cave is located near Mount Vesuvius, a volcano. Underground volcanic activity releases carbon dioxide, a gas that is heavier than air and displaces air to a height of a couple of feet inside the cave. A man walking in the cave will not be affected, but a dog will find no oxygen to breathe and will suffocate. Obviously, it is not a good idea for anyone to lie down to take a nap in the Grotto del Cane. It will be one long nap.

♔

Why did European pharmacies in the nineteenth century start selling powdered chrysanthemum flowers?

They were to be used as insecticides to keep bugs from invading gardens. Over two thousand years ago the Chinese discovered that the oil produced by a species of chrysanthemum was toxic to insects. The active ingredients have now been isolated, and the mixture of the six naturally occurring insecticidal compounds found in the head of the chrysanthemum flowers is referred to as pyrethrum. But long before the active ingredients were identified, Napoleon's armies were already using the powdered flowers to control lice infestations. Today, chrysanthemums are commercially grown, and the insecticidal oil is extracted and commonly used in lice treatments, mosquito repellents and organic agriculture.

Pyrethrum breaks down quickly when exposed to light, so chemists wondered whether compounds that differed only slightly in molecular structure would retain the insecticidal properties and last longer in the field. They ended up synthesizing a variety of pyrethroids, which are now commonly used. Permethrin, for example, is sprayed on cotton, wheat and corn and is also used to kill parasites on chickens. Like all insecticides, pyrethrum and pyrethroids are not risk free. They tend to be non-specific and destroy beneficial insects as well as pests. The U.S. Environmental Protection Agency has labelled pyrethrum and its analogues as possible human carcinogens based on animal studies. This of course does not mean that these chemicals cannot be safely used. Like synthetic pesticides, when used according to label directions, the risks are minimal.

Every day, each one of us consumes about a hundred times of our own weight of this substance. What is it?

Water. A little bit of a trick question because consumption in this case obviously cannot mean only oral consumption. But we are indeed responsible for the consumption of massive amounts of water because virtually everything we eat, everything we wear and every item we use requires water in its production. For food, the amounts are staggering. Would you believe that it can take up to 1,300 gallons (5,000 litres) of water to grow 2 pounds (1 kilogram) of rice, or 3,000 gallons (11,000 litres) to produce just one hamburger? This of course refers to the water needed to grow the feed for the cow that ends up as hamburger. But coffee takes the cake. It requires 5,300 gallons (20,000 litres) of water to produce 2 pounds (1 kilogram) of coffee. And that's not all. You could fill twenty-five bathtubs with the water needed to grow the cotton needed to make one T-shirt. But I know what you're thinking. Sure we use all that water, but we eventually get it back. Didn't we learn in school about the water cycle? We water the plants, the water is used to form the building blocks of the plants, but when we or animals eat the plants, the components are eventually metabolized, and one of the metabolic products always is water. So we produce water, which eventually ends up back in the environment. That is true enough.

The overall water content of the earth is constant. But the problem is that the water does not end up where it came from. Consider the situation in India. There is not enough surface water for irrigation, so farmers sink long tubes into the ground and pump water with electric pumps to the surface. The crops they grow and the milk they produce will eventually give back the water, somewhere. But this water will not end up as rain to replenish the original groundwater the Indian farmers used. Whereas before wells dug to 30 feet (10 metres) yielded water, now farmers sometimes have to dig down to 1,300 feet (400 metres). The long-term consequences are very serious, with populations facing starvation resulting from water shortage. Efforts are under way in India to capture rainwater as efficiently as possible, but this is unlikely to solve the problem. When we talk about an endangered food supply, we usually worry about depleted soil and losses to insects, fungi, weeds and the weather. But ultimately we may face a global food crisis for lack of water!

$\mathcal{Q}$

Why did Cleopatra supposedly bathe in sour donkey milk?

To improve the appearance of her skin by reducing wrinkles. When milk sours, bacteria convert the milk sugar lactose into lactic acid. When alpha hydroxy acids such as lactic acid are applied to the skin, they cause the surface layer to peel off, leaving new, smoother, blemish-free skin underneath. It is questionable whether sour milk has enough lactic acid to rejuvenate the skin, but modern cosmetics that contain at least 8 percent alpha hydroxy acids can reduce minor wrinkles. A truly effective chemical peel, though, requires the application of irritants such as trichloroacetic acid or phenol to the skin by a physician. The corrosive effect causes burning and stinging for several minutes followed by reddening and peeling of the skin over the next few days. As this happens new skin with a more youthful appearance forms. Chemical peeling is not a pleasant procedure, and the unsightly scab formation can be psychologically disturbing. Most patients are not ready for public viewing for several weeks, but in the end they tend to be satisfied with the results.

Cleopatra was apparently into more than just sour milk. She used powdered excrement from crocodiles to embellish her complexion, although this was in all likelihood fruitless. Perhaps the concoction did trigger Cleo's interest in perfumes. In her perfume factory, yes, she apparently was a venture capitalist. Herbs, flower petals, leaves or seeds were mixed with hot vegetable oil made from pressed olives. The mixture was allowed to soak for a week and then was pressed through a cloth bag to extract the perfumed oil. The queen even played around with baldness remedies, spurred by her relationship with Julius Caesar, who was hair challenged. She experimented with a goo made of ground horse teeth and deer marrow to coax his dormant hair follicles into action. When this didn't work Cleo traded Julius in for Mark Antony. And those beautiful eyes seen in ancient depictions of the famed Egyptian queen? Made up with green copper malachite and black lead sulphide. Not only did these improve her appearance but the chemicals kept flies away.

☙

A bee automatically sticks out its tongue when it expects food. Why are researchers at the Pentagon interested in this behaviour?

The idea is to use bees to detect explosives. Bees have a very keen sense of smell, and like dogs can detect odours at the part-per-trillion level. That would be roughly one small drop of perfume in an Olympic-size swimming pool. And they are easier to train than dogs. All you have to do is expose the insects to an odour while offering them a treat of sugar water. After just four short exposures of a few seconds each, the bees learn to associate the smell with food, and next time they are exposed to the scent they will exhibit a "proboscis extension reflex," meaning that they will stick out their tongue. They can recognize virtually any scent, be it that of ripe strawberries, drugs or explosives. A British company, Inscentinel, has actually built a prototype detector using three bees secured in tiny containers with only their heads sticking out. A miniature camera records the tongue behaviour as a fan draws air past the insects' heads, relaying the images to a laptop. The little detectives have already shown they can identify explosive vapours and are being put to a test at a British airport.

Wasps are also great at detecting odours. Some lay their eggs inside live caterpillars; they zero in on their hosts using the alarm smells plants release when they are attacked by caterpillars. Like bees, wasps can be trained with sugar water to detect the scent of TNT, or of cadaverine and putrescine given off by decomposing bodies, or 3-octanone given off by fungi. The response of wasps to a scent is different from that of bees. Females will coil in response to caterpillar-generated plant smells, and will drop their antennae to the ground when responding to food fragrances. Confining the insects in a box equipped with a video camera can allow technicians to monitor the wasps' behaviour and identify the presence of compounds undetectable by the human nose. In the future, such devices may even be used to detect disease. Ulcers, some cancers and tuberculosis have signature odours, and may be detected by insects before symptoms appear.

Insects are not the only creatures being investigated for their ability to detect smells. Believe it or not, giant pouched rats have been trained to sniff out landmines and to pick up the scent of tuberculosis in saliva. Even more fascinating is the possibility of using fish to detect tampering with drinking water. Bluegill sunfish cough when toxins are present in the water. Electrodes planted in the fish respond to this behaviour, sending an electronic signal to a computer. A New York reservoir is currently trying out the fish detection system.

♀

In 1919 an explosion in Boston released what sort of gooey material that killed 21 people and injured 150 others?

Molasses. A giant tank holding close to 2.5 million gallons (9.5 million litres) of molasses exploded, spewing chunks of metal all over and flooding the area with the black sticky stuff. People, animals and even homes were carried away. The tank belonged to the United States Alcohol Company, with the molasses destined to be fermented into rum. Actually it is the sugar content of molasses that is the raw material for fermentation. When sugar cane is processed, it is crushed and boiled to produce a thick slurry from which sugar crystallizes out. The syrup left behind after filtering the sugar crystals still contains lots of dissolved sugar and is called molasses. Why the tank ruptured that day isn't clear, but a large temperature change was probably instrumental. A sudden warming trend in January caused the temperature to rise from about 2°F (-15°C) to 40°F (4°C), causing the molasses to expand and rupture the tank. The distillery was brought to trial and ended up paying out more than a million dollars in damages. More than six months were needed to remove the molasses from the streets, and apparently Boston harbour was stained brown for months.

How can crime scene investigators determine whether a light bulb was on or off when it was broken?

By examining the filament and establishing whether it is coated with tungsten oxide, a yellowish white substance. The filament inside an incandescent light bulb is made of tungsten, a metal that glows red hot when an electric current is passed through it. At high temperatures tungsten readily reacts with oxygen to form tungsten oxide, which is why light bulbs are filled with an inert gas such as argon instead of with air. If a bulb is broken when it is on, the hot tungsten will immediately react with oxygen and show a deposit of tungsten oxide. If the bulb was cool, there will be no such deposit. Crime scene investigators would be interested in determining whether a bulb was on or not in the case of a car accident in which there is a question of whether the headlights were on when a crash occurred.

Although tungsten does not react with argon, light bulbs still have a fixed lifetime. That's because each time the filament gets hot, some of the tungsten evaporates, making the filament thinner and thinner until it eventually breaks. And what happens to the tungsten that evaporates? It deposits on the cooler surface of the glass, forming a dark spot. Such deposits are readily visible on older light bulbs.

What crime involves the use of dead insects?

The manufacture of fake amber. Amber is the fossilized resin of trees that lived millions of years ago and is widely used today to make jewellery. The most valuable pieces are those that feature an entombed insect, trapped in the resin as it oozed from trees. Amber differs from other types of gems in several ways. Diamonds, for example, are composed only of carbon. Pearls are made of calcium carbonate. Emerald is essentially beryllium aluminum silicate with traces of chromium or vanadium that impart colour. Essentially, these gems are composed of simple inorganic substances. By contrast, amber is a complex mixture of organic compounds, the term being used here in its proper chemical sense. Organic compounds all contain carbon atoms joined in various ways. The resin released by trees contains hundreds of different compounds, some of which have insecticidal properties, others of which act as fungicides, while some join together in long chains to form a sticky substance that seals injuries to the bark. Terpenes are the components that can readily polymerize to harden the resin, trapping the other components.

Because of the increasing value of amber, forgers have entered the picture, often duplicating the appearance and feel of genuine amber, even managing to embed "inclusions." Several methods are used. Plastics such as Bakelite and celluloid can be used but can be readily detected by the "hot needle" test. Pressing a heated needle to the material immediately releases a burnt plastic smell, while real amber smells like pine trees. Fakes made with copal, a semifossilized resin, are harder to detect. Copal essentially is very young amber, if you can call anything that is at least a million years old young. All trees ooze resin, but it takes millions and millions of years for this to harden into amber. Copal has the appearance of amber but is softer. When heated it becomes sticky much more readily than true amber. The really ingenious crooks take a piece of real amber, slice off an end, drill a hole and place an insect inside. Then they melt the piece they sliced off and pour it over the hole. When it hardens, you have a genuine piece of amber with an insect inside. And money in your pocket. The newly created gem can be sold as an item of jewellery, or in some truly fraudulent instances, as a healing substance.

Amber has a long tradition of supposed healing, with Hippocrates recommending it for delirium and poor eyesight. Martin Luther always carried a piece for protection against kidney stones. There is no record of his ever suffering from such, so he probably concluded it worked. Today, some alternative health practitioners claim various healing properties for amber. One thing about fake amber, it is likely as effective at healing as the genuine stuff.

Amber has, however, made a real scientific connection. Over 2,500 years ago, Thales of Miletos discovered that when amber was rubbed against cloth, sparks were produced and then the amber attracted husks and small wooden splinters. This force was given the name electricity, after the Greek word *electron*, which means "amber."

curiouser and
curiouser

To what does legendary Chicago bartender Mickey Finn owe his fame?

Dropping chloral hydrate, a sedative, into his patrons' drinks. The idea was not to settle down unruly customers but to knock them out so that they could be deprived of some of their possessions. Chloral hydrate was first made by Justus von Liebig, the brilliant German chemist, in 1832, but it was only in 1869 that its sedative and sleep-inducing properties were discovered. Indeed, it is recognized as the first commonly used drug to induce sleep. Mickey Finn supposedly opened his bar in Chicago in 1896 and within two years had become adept at using his "knockout drops." Chloral hydrate is still available today but to a large extent has been supplanted by the benzodiazepine sedatives like Valium and Librium. Since it may be habit forming, its use is carefully regulated.

For what purpose would a magician use zinc stearate?

To prevent playing cards from sticking to each other. Card tricks are perhaps the most popular of all magic tricks. They often start with the magician spreading the cards with a flourish. Often with a little help from chemistry! A really nice card fan requires that the cards be lubricated. The classic choice is a fine white powder called zinc stearate. It is often used as a lubricant to prevent sticking in the moulding of plastics and is a common ingredient in face powders. It is also a component in many pills, where it serves as an internal lubricant, enhancing the rate at which the pill disintegrates in the digestive tract. Magicians place a little zinc stearate in a paper bag and shake the cards with it. It is virtually imperceptible once applied and allows for smoother card handling.

♀

On April Fool's Day in 2000, Britain's *Daily Mail* newspaper ran a story about a revolutionary new line of socks. What were these socks supposed to do?

Help people lose weight. FatSox were supposedly the invention of Professor Frank Ellis Elgood and were made of a patented polymer called FloraAstraTetrazine "that had previously only been applied in the nutrition industry." FatSox, it was said, could banish fat forever. How? The special fabric sucked fat out of the feet as a person sweated. As body temperature rose, blood vessels dilated and then the socks "drew excess lipid from the body through the sweat." The socks could then be washed to remove the fat and could be used again. Needless to say, many consumers contacted the paper to find out where the miraculous FatSox could be purchased.

Of course, FatSox never existed, but equally ridiculous products do exist and are widely promoted. Electrical muscle stimulators that claim to "burn fat while you sleep" and reducing pyjamas that melt fat away are available along with magnetic belts guaranteed to get rid of those extra pounds. Or how about the Acu-Stop 2000? This was a rubber device that was to be inserted into the ear and was designed to align with the "pressure points that control appetite." The user was advised to insert it into the ear several times a day and massage it for several minutes. It sold for $39.95 and cost roughly 17 cents to produce. Then there was Phena-Drene, a pill that was supposed to turn fat into water, which would flow right out of the body by the gallon. And if none of those worked, you could always try "magic weight loss earrings" or "appetite-suppressing eyeglasses." These may actually work if they help you see yourself better in the mirror.

❦

Vladimir Lenin lies in a mausoleum in Red Square in Moscow in a preserved state. What part of his body is missing?

The brain. That's because the Soviet government had it removed after he died in 1924. Lenin's brain was to be examined scientifically to see how it differed from that of mere mortals. After all, someone who thought up something as clever as communism must have had a very special brain. The item in question was sent to Oskar Vogt, a well-known German neuroscientist, who spent a couple of years studying it. His conclusion? At least one part of the cerebral cortex had more and larger nerve cells than expected. Perhaps it was this area of the cerebral cortex that was exercised by thoughts of communism. Lenin's brain was not the only one studied in search of physical signs of great intellect. Perhaps the most famous brain in the world belonged to Albert Einstein. Al died in 1955 and was cremated. But his son thought that such a wonderful brain was too hot an item to burn and decided it should be saved for research. An autopsy was performed on Einstein by Dr. Thomas Harvey, who removed the brain.

The brain was sectioned and examined, but the findings were not made public. Eventually the pieces were packed in two jars, which Dr. Harvey kept as a bizarre souvenir. Sometime in the 1970s a reporter got wind of the Einstein brain business, tracked down Dr. Harvey and wrote an article on the weird affair. Dr. Harvey began to receive all sorts of requests for samples, some of which he did provide to respected neuroscientists. Eventually three scientific papers were published about the brain, each one finding some differences between it and ordinary brains. There was nothing really dramatic, except for the fact that some parts of the brain had a greater density of nerve cells. Einstein's brain weighed less than that of the average adult male, so certainly, in this part of the body, it is not size that counts.

Small or not, Einstein's brain was spectacular. He always had a brilliant answer to a question. When he was queried about the concept of relativity he had a brilliant answer. "Put your hand on a hot stove for a minute, and it seems like an hour. Sit with a pretty girl for an hour, and it seems like a minute. That is relativity!"

The musical *Chicago* features a song with the lyrics "You can look right through me, walk right by me, and never know I'm there." What material is the singer referring to?

Cellophane. The song is "Mr. Cellophane" and humorously describes the fear of being inconsequential. Well, cellophane may be clear, and almost invisible, but it certainly is not inconsequential. The name of this material derives from *cellulose* and the French word *diaphane*, meaning "transparent." Cellophane of course is a transparent film that is used to wrap almost anything you want and is also the essence of Scotch tape. Like so many discoveries, it came about in an accidental fashion. According to the story, which may even be true, Swiss textile engineer Jacques Brandenberger was dining in a restaurant when a customer spilled some red wine on a tablecloth. As Brandenberger watched the waiter's desperate efforts to deal with the problem, he wondered if there were a way to coat tablecloths to make them resistant to stains. His mind flipped to a material called viscose that the textile industry was already very interested in because of its potential to produce fibres. Viscose was made by treating raw cellulose with caustic soda and carbon disulphide.

Brandenberger wondered if somehow the material could be used to cover fabric. He tried and tried, with no success. The thin film of viscose he applied to cloth kept peeling off. Not only did it peel off, it did so in a flexible, transparent sheet. Now the Swiss scientist shifted focus and decided that the film itself could be useful. Indeed it was! He immediately thought of cellophane as a replacement for flammable celluloid in movie film, but cellophane distorted under heat, so it was unsuitable for this purpose. As a packaging material, though, it was great. The first customer was Coty, the French parfumeur, which used it to wrap perfume bottles.

It was also used to wrap chocolates, but here a problem cropped up. Under moist conditions, cellophane became sticky. On the other hand, it was impermeable to poison gases, and in the First World War it found extensive use as a lens in gas masks. The moisture problem was solved in the 1920s by William Hale Charch, a chemist working for DuPont who had bought the rights to cellophane. He found that a thin coating of nitrocellulose made cellophane waterproof. This was just what the makers of Camel cigarettes needed. They hatched an advertising scheme promoting the fact that they were now packaging their cigarettes in "humidor." Cellophane of course is still with us. An American poll in 1940 searched for the most beautiful word in the English language. *Cellophane* placed third, behind *mother* and *memory*.

♀

In 2002, the Royal Society of Chemistry in Britain bestowed an Extraordinary Honorary Fellowship on a man who never lived. Who was it?

Sherlock Holmes. Sir Arthur Conan Doyle's fictional detective had a "profound knowledge of chemistry," as his chronicler, Dr. John Watson, informs us in *A Study in Scarlet*. And he so often put this knowledge to good use that the Royal Society decided Holmes was worthy of an Honorary Fellowship. In "The Adventure of the Naval Treaty," Watson gives us a glimpse into Holmes's chemical investigations. "Holmes was seated at his side-table clad in his dressing-gown and working hard over a chemical investigation. A large curved retort was boiling furiously in the bluish flame of a Bunsen burner, and the distilled drops were condensing into a two-litre measure. My friend hardly glanced up as I entered, and I, see-ing that his investigation must be of importance, seated myself in an arm-chair and waited. He dipped into this bottle or that, draw-ing out a few drops of each with his glass pipette, and finally brought a test-tube containing a solution over to the table. In his right hand he held a slip of litmus-paper. 'You come at a crisis, Watson,' said he. 'If this paper remains blue, all is well. If it turns red, it means a man's life.' He dipped it into the test-tube, and it flushed at once into a dull, dirty crimson. 'Hum! I thought as much!' he cried. . . . 'A very commonplace little murder,' said he."

We are left up in the air about just what sort of investigation the great detective was pursuing, but it is likely to have had something to do with poisons, a subject that certainly interested him. In "The Adventure of the Devil's Foot," for example, Holmes figures out that a murder was committed with a volatile toxin extracted from the root of an African plant. While the devil's foot root is fictitious, plant toxins are not. In fact Conan Doyle earlier in his career as a physician wrote about gelseminum, a potentially poisonous prepa-ration from the root of the yellow jasmine. In the case of the Sussex Vampire, Holmes uses his knowledge about South American arrow poisons to clear an unfortunate lady wrongly accused of vampirism and fingers the real criminal. Science triumphs over nonsense. As Holmes himself underlines, "This agency stands flat-footed upon the ground, and there it must remain. The world is big enough for us. No ghosts need apply." An interesting quote, given that Conan Doyle, Holmes's creator, believed in ghost and spirits. Although Holmes never lived, his spirit, which will live on forever, is certainly worthy of the Royal Society of Chemistry's Honorary Fellowship.

How can you tell if an ice cube has been made from tap water or from distilled water?

Ice cubes made from distilled water are clear, whereas those made from tap water are cloudy. Distilled water is made by boiling water, converting it into water vapour and condensing the vapour back to a liquid in a separate container. Tap water contains a variety of minerals, mostly salts of calcium and magnesium. When the water is distilled, these minerals are left behind. When tap water freezes, the liquid solidifies and the minerals are pushed out of solution. These mineral refract light and create an opaque ice cube.

☙

A U.S. patent has been issued for a toy that makes use of human flatus. What is the toy?

A rocket. As incredible as it sounds, U.S. patent 6,055,910 has been issued to Michael Zanakis and Philip Femano for a "toy gas-fired missile and launcher assembly" that is powered by human flatus. The invention describes how the operator places an inlet tube equipped with a valve adjacent to his anal region to capture the colonic gas, which is then directly introduced into the combustion chamber of the rocket. An "igniter is then activated to explode the gas and fire the missile into space." The patent has an elaborate description of the device, replete with drawings. But there is no evidence that such a rocket has ever been built, let alone flown. In order to get a patent, all one has to do is demonstrate that an invention is novel, not that it actually works. Zanakis and Femano offer a rationale for their invention in the patent: "A recreational activity practiced by some individuals is ignition of one's own flatus. This is performed by using a lit match or candle, or a cigarette lighter. So widespread is this activity that there are web sites on the Internet devoted exclusively to explaining proper lighting techniques. A major drawback of this popular practice is that it usually involves the hazardous coupling of fire, combustible gases and inebriated participants. Reports of serious burns to body parts are not uncommon, this being especially true when the participants remove their clothing." So it seems the inventors' motive is to provide a safe recreational flatus activity.

Could the device actually work? All rockets make use of Newton's third law, namely that for every action there is an equal and opposite reaction. Combustion processes that expel gases through the exhaust of a rocket propel the vehicle in the opposite direction. And certainly human effluvia do contain some combustible gases, such as methane and hydrogen. These, however, do not make up the bulk of the released gas; nitrogen, oxygen and carbon dioxide are more prevalent. How much methane and hydrogen are produced depends on a variety of factors, among them the composition of the diet, genetics and the presence of bacteria in the colon. Certain carbohydrates in food, such as stachyose and raffinose in legumes, are fermented by colonic bacteria and yield combustible gases. But even if a subject loaded up on beans, the amount of gas that could be captured via the method suggested in this invention would not be enough to power a rocket. This silliness does serve a purpose, though. It shows that a claim that something is patented does not mean that it works effectively. This goes for dietary supplements, pain-control devices, transparent globes for transcendental meditation—and anal rockets.

Chewing gum containing phytoestrogens from the
Pueraria mirifica plant comes with a claim of what
sort of biological effect?

That it can increase bust size. Bust Up gum is promoted on
numerous websites as a "natural" way to increase breast size and is
enjoying brisk sales, especially in Japan. The concept behind this
product is actually legitimate, since substances that mimic the
body's estrogen can have an effect on the size of the breasts. And
Pueraria mirifica, which grows in Thailand and Burma, does contain
miroestrol, deoxymiroestrol and isoflavones, all of which have
estrogenic activity. Marketers claim that chewing the gum three or
four times a day will increase breast size in over 90 percent of
women, but no published studies yet back up this claim. Neither
do any studies demonstrate the safety of the product. Altering
estrogen levels is not risk free. Given that studies have shown a link
between hormone replacement therapy, which makes use of estrogenic substances, and breast cancer, caution with any type of estrogen is appropriate. Contrary to what the advertising implies, the
fact that the estrogens derive from a "natural" plant source, does
not absolve these substances of possible mischief.

This is not the first time that chewing gum has been promoted
with some sort of health claim. In 1869, dentist William Semple
patented chewing gum as a jaw exerciser, although it was never clear
why the jaw needed to be exercised. More recently, it has been
claimed that Brain Gum increases mental alertness because it contains phosphatidylserine, a compound found in brain cells. It
comes replete with some not too convincing studies about improving memory. Still, there may be something to the idea that chewing
gum can make us smarter. A British study tested short-term memory such as recalling words, pictures and phone numbers in volunteers who chewed gum, pretended to chew gum or did not chew

gum. For three minutes they chewed, pretended to chew or did not chew, and then partook in the memory test. Gum chewers performed better than others. Researchers theorize that chewing may stimulate insulin release, and insulin enhances brain activity. Or that a slight increase in heart rate associated with chewing causes more oxygen to be delivered to the brain. So maybe those women fervently chewing Bust Up gum will increase their brain power enough to realize that chewing the gum is likely to be a bust.

The chemical composition of the ice was changed for the 1949–1950 National Hockey League season. What change was introduced?

The ice was painted white. Artificial ice, which is a misnomer since it is very real ice, is made by pouring water over a concrete surface that can be cooled by circulating refrigerated brine solution through pipes embedded in it. The colour of the concrete can be seen through the ice, which appears a dull grey, providing less than ideal contrast with the black puck. At the beginning of the 1949–1950 season the National Hockey League decided to paint the ice to help spectators to see the puck more clearly. The white pigment used was probably titanium dioxide, although at the time lead carbonate was still a commonly used pigment and could have been a component of the paint. White light is a mixture of all visible wavelengths, and these pigments reflect all wavelengths. Colour appears when some of the wavelengths are absorbed by a material and others are reflected. An ideal white paint will reflect all wavelengths. Today, titanium dioxide, zinc oxide and calcium carbonate are the pigments used in white paints, including the ones specially formulated for ice. Television viewers got their first glimpse of painted ice on October 11, 1952, with the initial *Hockey Night in Canada* telecast. When colour television came on the scene the brilliant white ice produced too much glare and the surface had to be painted blue. Technology soon overcame this faux pas so that today we can watch hockey played on a white sheet of ice, as it should be.

Real "artificial ice" does exist, made of polymers treated with a lubricant. Ice skates can be used of course, but you can't get that great spray of ice particles when stopping. Interestingly, these artificial surfaces are also treated with titanium dioxide to make them white.

♀

What single word does the code "Answer, tell, pray, answer, look, tell, answer, answer, tell" represent?

"Believe." Houdini had planned to relay this code to his wife, Bess, from beyond the grave, if indeed such a communication were possible. The code was one Houdini and Bess used when they were performing a mind-reading trick early in Houdini's stage career. Harry Houdini dominated the world of entertainment in the early part of the twentieth century. He walked through walls, made elephants vanish and escaped from straitjackets, jail cells and handcuffs of every variety. He introduced the idea of the "challenge," whereby he invited audience members to come on stage and secure him with ropes and handcuffs of their choosing, from which he would invariably escape.

But there was another type of escape that Houdini eventually became interested in. An escape from the grave. Not the kind of buried-alive escape he had often performed for audiences, but a literal cheating of death. Harry became focused on this idea after the death of his beloved mother, when out of desperation he had visited a number of mediums in an attempt to contact her in the great beyond. Instead of comforting messages from his mother, Houdini found extensive fraud. Mediums were using the same kind of magic effects with which he had thrilled audiences, but they were using them to convince the gullible that they were making contact with the spirit world.

From then on Houdini devoted his career to unmasking fraudulent mediums, but he never completely ruled out the possibility of life after death. If anybody could escape the Grim Reaper, it would be he! Harry and Bess agreed on a message that if relayed by a medium after Harry's death would prove that contact had been made. There is controversy about whether this actually happened. In 1928 famed medium Arthur Ford told Bess that while in a trance he had received the message "Rosabelle, answer, tell, pray, answer, look, tell, answer, answer, tell." The code was indeed the one that Harry and Bess had agreed upon, and Rosabelle was the name etched into Bess's wedding ring because it was a song Bess liked when the two had initially met. There are all kinds of theories floating around about how Ford may have learned of the code, even that Bess had revealed it to some friends earlier. In any case, Bess kept on trying to contact Harry through the famed Halloween Eve séances for ten years, when after another failure she finally gave up, saying, "My last hope is gone. I do not believe that Houdini can come back to me or to anyone. The Houdini shrine has burned for ten years. I now, reverently, turn out the light. It is finished. Good night, Harry." It seems that ten years was long enough to wait for any man.

index